AF251028

Preparation and Properties of Solid State Materials

VOLUME 6

Lead Tin Telluride, Silver Halides and Czochralski Growth

PREPARATION AND PROPERTIES
OF SOLID STATE MATERIALS

Editor

W. R. WILCOX

Department of Chemical Engineering
Clarkson College of Technology
Potsdam, New York

Associate Editors

S. B. Austerman
Rockwell International
Anaheim, California

J. O'M. Bockris
School of Physical Science
Flinders University of
   South Australia
Bedford Park, South Australia

John Carruthers
Director, Materials Processing
   in Space
NASA Headquarters, EP
Washington, D.C.

R. H. Doremus
Materials Division,
School of Engineering
Rensselaer Polytechnic Institute
Troy, New York

S. Fujiwara
Institute for Optical Research
Kyoiku University
Tokyo, Japan

M. S. Larson
Department of Chemical Engineering
   and Nuclear Engineering
Iowa State University
Ames, Iowa

R. A. Laudise
Bell Laboratories
Murray Hill, New Jersey

R. A. Lefever
Ampex Corporation
El Segundo, California

T. B. Reed
MIT Lincoln Laboratory
Cambridge, Massachusetts

D. W. Shaw
Central Research Laboratories
Texas Instruments, Inc.
Dallas, Texas

R. M. Spriggs
Lehigh University
Bethlehem, Pennsylvania

E. A. D. White
Alhazen Research Institute
Baghdad, Iraq

A. Wold
Department of Chemistry
Brown University
Providence, Rhode Island

Preparation and Properties of Solid State Materials

VOLUME 6

Lead Tin Telluride, Silver Halides and Czochralski Growth

Edited by

WILLIAM R. WILCOX

Department of Chemical Engineering
Clarkson College of Technology
Potsdam, New York

MARCEL DEKKER, INC.   New York and Basel

Library of Congress Cataloging in Publication Data
Main entry under title:

Lead tin telluride, silver halides, and Czochralski growth.

   (Preparation and properties of solid state materials ; v. 6)
   Includes bibliographies and indexes.
   1. Crystals—Growth. 2. Lead tin telluride crystals.
3. Silver halide crystals. I. Wilcox, William R.
II. Series.
TA401.P655 vol. 6 [QD921] 620.1'1s [669'.4] 81-12584
ISBN 0-8247-1367-2   AACR2

COPYRIGHT © 1981 by MARCEL DEKKER, INC.   ALL RIGHTS RESERVED

Neither this book nor any part may be reproduced or transmitted in any form
or by any means, electronic or mechanical, including photocopying, micro-
filming, and recording, or by any information storage and retrieval system,
without permission in writing from the publisher.

MARCEL DEKKER, INC.
270 Madison Avenue, New York, New York 10016

Current printing (last digit):
10 9 8 7 6 5 4 3 2 1

PRINTED IN THE UNITED STATES OF AMERICA

PREFACE

This volume, Volume 6 of the series Preparation and Properties of Solid State Materials, provides practical advice for the crystal grower. Chapter 1 is a treatise on lead tin telluride, which has important applications for infrared sensing devices and lasers. Preparation, properties, and devices are all discussed, as well as the relationship between these. Although Chapter 2 covers precipitation of silver halides for photographic applications, the techniques and principles discussed are useful in other precipitations as well. Chapter 3 is a thorough treatment of heat transfer in Czochralski growth, including convection, interface shape, interfacial temperature gradient, diameter control, and doping.

William R. Wilcox

CONTRIBUTORS

ROWLAND E. JOHNSON[*], Physical Sciences Research Laboratory, Texas Instruments Incorporated, Dallas, Texas

NOBUYUKI KOBAYASHI, Department of Applied Physics, Faculty of Engineering, Toyama University, Takaoka, Toyama, Japan

SIDNEY G. PARKER, Physical Sciences Research Laboratory, Texas Instruments Incorporated, Dallas, Texas

JONG S. WEY, Research Laboratories, Eastman Kodak Company, Rochester, New York

---

[*] Present address: Middleton Road, Dallas, Texas.

CONTENTS

Preparation and Properties of Solid State Materials

VOLUME 6

Lead Tin Telluride, Silver Halides and Czochralski Growth

Chapter 1

PREPARATION AND PROPERTIES OF (Pb,Sn)Te

Sidney G. Parker and Rowland E. Johnson[*]

Physical Sciences Research Laboratory
Texas Instruments Incorporated
Dallas, Texas

(Pb,Sn)Te is a semiconductor which can be prepared to a high degree of crystalline perfection in large single crystals by vapor transport techniques. Epitaxial films of high perfection can be grown by either vapor-phase or liquid-phase methods. Because of the thermodynamic properties of the PbTe-SnTe system, (Pb,Sn)Te as prepared is usually nonstoichiometric with a large concentration of p-type defects. The carrier defect concentration can be varied by selection of growth

*Present address: Middleton Road, Dallas, Texas

temperature, annealing procedures, change of composition, or incorporation of electrically active dopants. The bandgap of $Pb_{1-x}Sn_xTe$ is direct but varies with composition, x, and temperature. The change in bandgap with composition has been explained by a band inversion model. The availability of high-quality (Pb,Sn)Te with its small direct bandgap has led to application studies of detectors and lasers. Wavelength tuning in lasers is possible by varying the composition, by varying the current, or by varying the temperature. Power output of the lasers is related to the number of dislocations present, to the geometry of the laser cavity, to the carrier concentration, and to the operating temperature. $Pb_{1-x}Sn_xTe$ lasers have been used for spectroscopic studies of various compounds and for the detection of numerous pollutants. $Pb_{1-x}Sn_xTe$ detectors cover the range from 2 to 20 μm and have shown excellent response at temperatures up to 200 K.

I. INTRODUCTION

Lead tin telluride, $Pb_{1-x}Sn_xTe$, is a semiconductor exhibiting many properties of intrinsic interest as well as properties applicable to device use. PbTe and SnTe are completely miscible so all values of x between 0 and 1 are accessible. Lead tin telluride crystallizes in a simple rock salt (cubic) structure which generally obeys Vegard's law. $Pb_{1-x}Sn_xTe$ has a direct bandgap which decreases linearly as x increases from zero; the bandgap passes through zero at $x = 0.35$ and increases linearly as x increases further. In addition, both n- and p-type materials are possible, produced by extrinsic doping or by self-doping with excess metal or tellurium atoms.

All the usual techniques of crystal growth have been used to grow crystals or films of (Pb,Sn)Te. This includes Czochralski, Bridgman, vapor transport, liquid-phase epitaxy, sputtering, and molecular beam methods. It has proven very difficult to grow single crystals free of imperfections, and the single crystals after growth are very susceptible to mechanical damage so special care in handling is necessary. Single-crystal films of (Pb,Sn)Te have been grown by vapor or liquid epitaxy on substrates such as BaF_2, Ge, NaCl, PbTe, and (Pb,Sn)Te.

As-grown (Pb,Sn)Te is usually nonstiochiometric because of vacancies. Material grown at or near the melting point is p-type with 10^{18} to 10^{19} cm^{-3} carriers due to metal vacancies; the carrier concentration is higher in material with a higher fraction of Sn. The p-type concentration can be lowered by annealing in the presence of a metal-rich (Pb,Sn)Te charge but several weeks is required to produce material with 10^{16} to 10^{17} cm^{-3} p-type carriers. Surface alloying and metal precipitation in the bulk are complications that occur during annealing. Recent advances in epitaxial growth have enabled several workers to obtain low carrier concentrations in either p- or n-type material by growing at temperatures well below the melting point. Doping during crystal growth or by diffusion into the grown material has been used to reduce the p-type carrier concentration, but these methods are frequently difficult to control.

Numerous models have been proposed for the band structure of (Pb,Sn)Te. These models were based on optical absorption, laser emission, and tunneling experiments and have been used to explain the variations in the bandgap with composition and with temperature. In addition, a number of studies have explored the unusual behavior of the bandgap, i.e., it shows a zero value at an intermediate Sn concentration and increases with both increasing and decreasing Sn concentrations from that composition.

One of the earliest applications of (Pb,Sn)Te was based on the thermoelectric properties of this material. More recently infrared detectors and lasers have been made from these materials. These devices usually function at temperatures $\leq$ 200 K. A more recent goal in detector research has been the fabrication of integrated systems, for example, a (Pb,Sn)Te detector array with a signal processor in a monolithic or hybrid structure for data reduction at the detector focal plane. Lasers made from (Pb,Sn)Te have been used for spectroscopic studies ot several compounds and offer a great potential for monitoring atmospheric pollutants.

II. PHASE EQUILIBRIA

Abrikosov and coworkers [1] were the first to report that PbTe and SnTe form a continuous series of solid solutions. Their data were based on thermal analysis and x-ray diffraction studies and did not account for variations in lattice parameter caused by nonstoichiometry. Their results were confirmed by others with more accurate and consistent values of the lattice parameter [2,3]. Work by Reti et al. [3] indicates that solid solutions of PbTe and SnTe approach ideal behavior:

> Small amount of heat liberated on mixing.
>
> Solid solution hardening is minor.
>
> Pb and Sn atoms are randomly distributed on their crystalline sublattices.
>
> The lattice parameter changes linearly with composition.

Figure 1 shows the temperature-composition phase diagram of the PbTe-SnTe system by Calawa et al. [4]. As indicated, only pseudobinary compounds exist in the solid with a relatively narrow separation of the solidus and liquidus lines.

As with PbTe and SnTe, $(Pb,Sn)Te$ exhibits nonstoichiometry caused by Pb and/or Sn vacancies or interstitials. However, the observed nonstoichiometry is very small, i.e., in $(Pb_{1-x}Sn_x)_{1-y}Te_y$, y lies very close to 0.5. Figure 2(a) shows the Pb-Te phase diagram from studies by Melngailis and Harman [5]. Figure 2(b) gives a greatly expanded view in the vicinity of the stoichiometric compound composition. This indicates that PbTe grown from a stoichiometric melt will be p-type due to excess Te. Figure 2(b) also shows that $(Pb,Sn)Te$ grown at the maximum liquidus temperature is shifted toward even greater p-type carrier concentrations. This concurs with the observation that melt-grown PbTe and SnTe contain $\sim 10^{18}$ and 10^{21} cm^{-3} p-type carriers, respectively. In later discussion, we show that the carrier type and concentration in $(Pb,Sn)Te$ depend on the composition, the temperature and method of preparation, and the thermal history of the material.

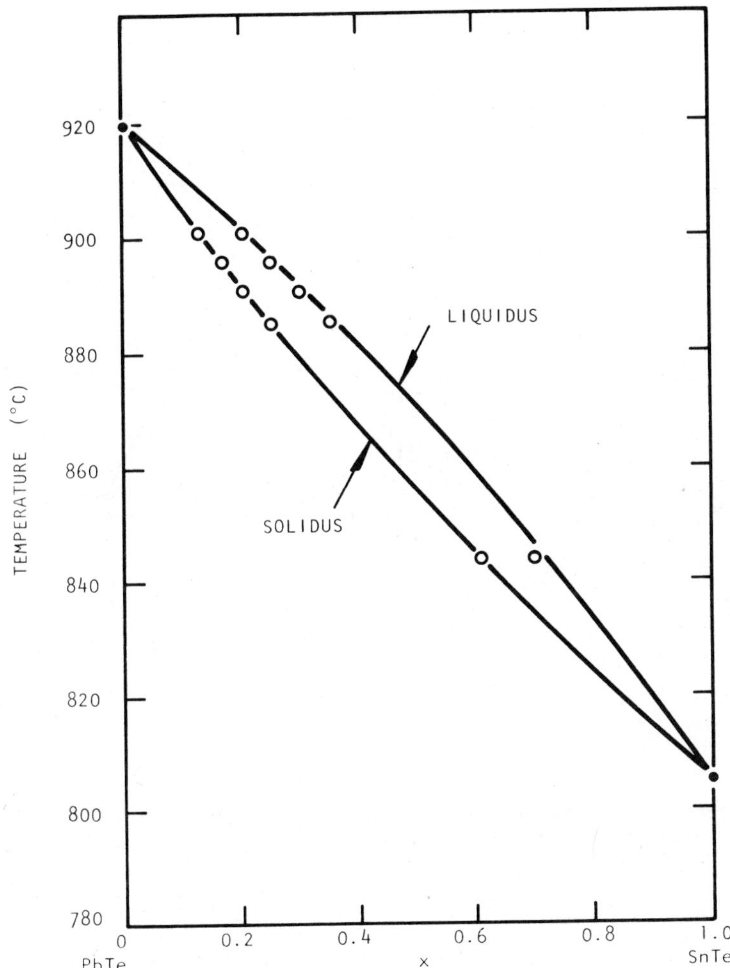

FIG. 1.   Temperature-composition phase diagram of the pseudobinary
$Pb_{1-x}Sn_xTe$ system.   [From A. R. Calawa, T. C. Harman, M. Finn, and
P. Youtz, *Trans. TMS-AIME*, *242*, 374 (1968); reprinted with permission.]

Laugier et al. [6] calculated the ternary-phase diagram for the
Pb-Sn-Te system assuming a regular associated solution.   The model
was based on equilibrium between the pseudobinary solid solution
(Pb,Sn)Te and the ternary liquid solution Pb-Sn-Te, where stable
complexes of PbTe and of SnTe also exist in solution in equilibrium

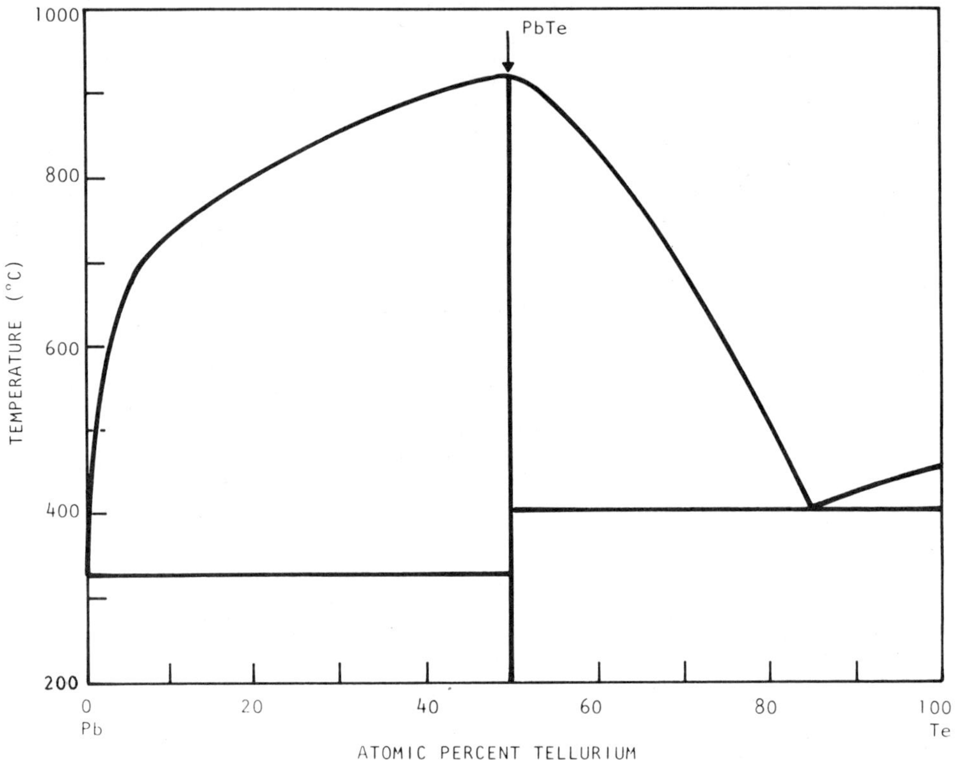

FIG. 2(a). Binary phase diagram for PbTe. (From Ref. 5; reprinted
with permission.)

with their respective elemental constituents. Calculated values
agreed well with experimental data on liquidus and solidus points
[6-8]. Figures 3(a) and 3(b) show the ternary liquidus isotherms and
isosolid composition lines respectively as calculated by Laugier et
al. and used by them to calculate liquid-phase epitaxial (LPE) growth
conditions.

Harris et al. [9] also calculated the ternary phase diagram for
(Pb,Sn)Te using a modification of a simple solution thermodynamic
model. Interaction parameters between elements in solution were determined by fitting a polynomial in temperature to experimental liquidus
and solidus data. The calculated liquidus and solidus lines for

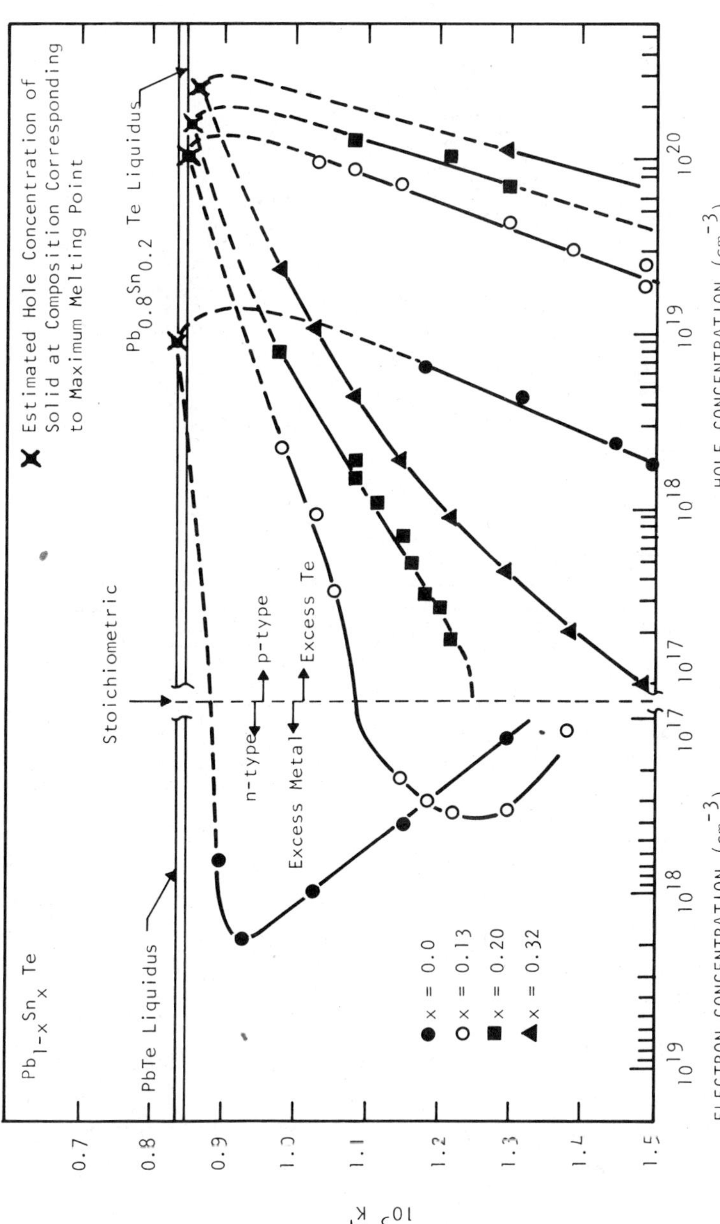

FIG. 2(b). Carrier concentration calculated from Hall coefficient measurements at 77 K as a function of isothermal annealing temperature for several Pb$_{1-x}$Sn$_x$Te alloy compositions. (From Ref. 51; reprinted with permission.)

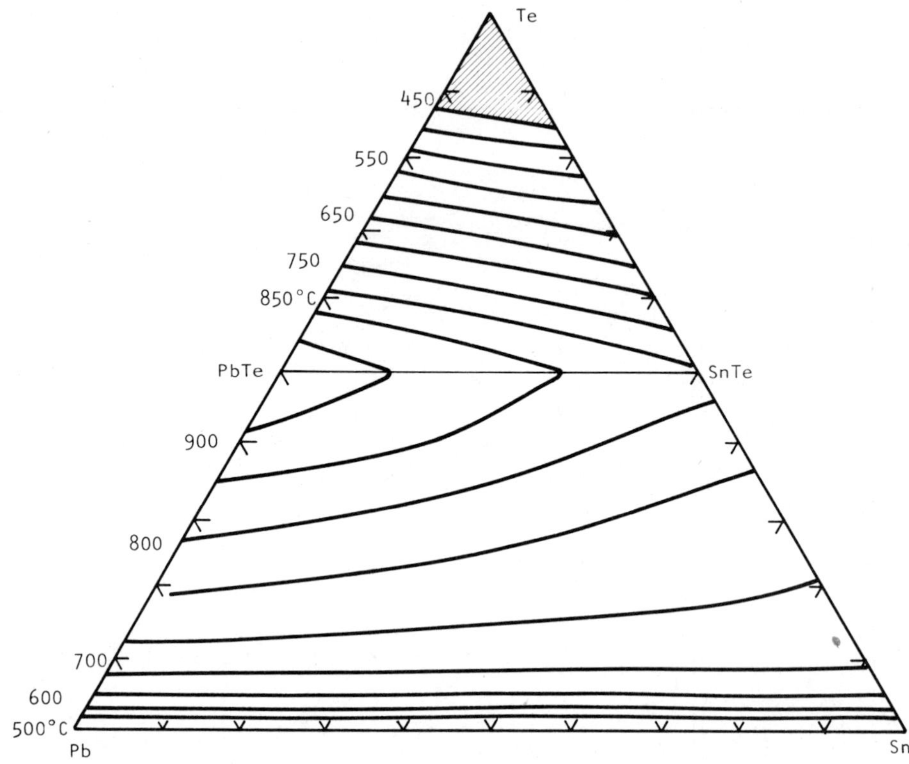

FIG. 3(a). Calculated ternary liquidus isotherms of Pb-Sn-Te. The
monotectic (dashed zone) is not considered. [From A. Laugier, J.
Cadoz, M. Faure, and M. Moulin, *J. Cryst. Growth*, *21*, 235 (1974);
reprinted with permission.]

Pb-Sn-Te are shown in Fig. 4. The calculated data of Harris et al.
gave a better fit to liquid-phase epitaxial data than those of Laugier
et al. Excellent agreement between calculated and experimental values
of liquidus and solidus was also obtained [2,10].

The compositional stability limits for $(Pb_{1-x}Sn_x)_{1-y}Te_y$ were
determined by Brebrick [11] using lattice parameter measurements at
25°C on powder samples for x = 1, 0.90, 0.80, and 0 and various values
of y, both within and outside the compositional stability range. Powdered samples of selected x were annealed in the presence of metal-
rich or tellurium-rich material to achieve a desired value of y.

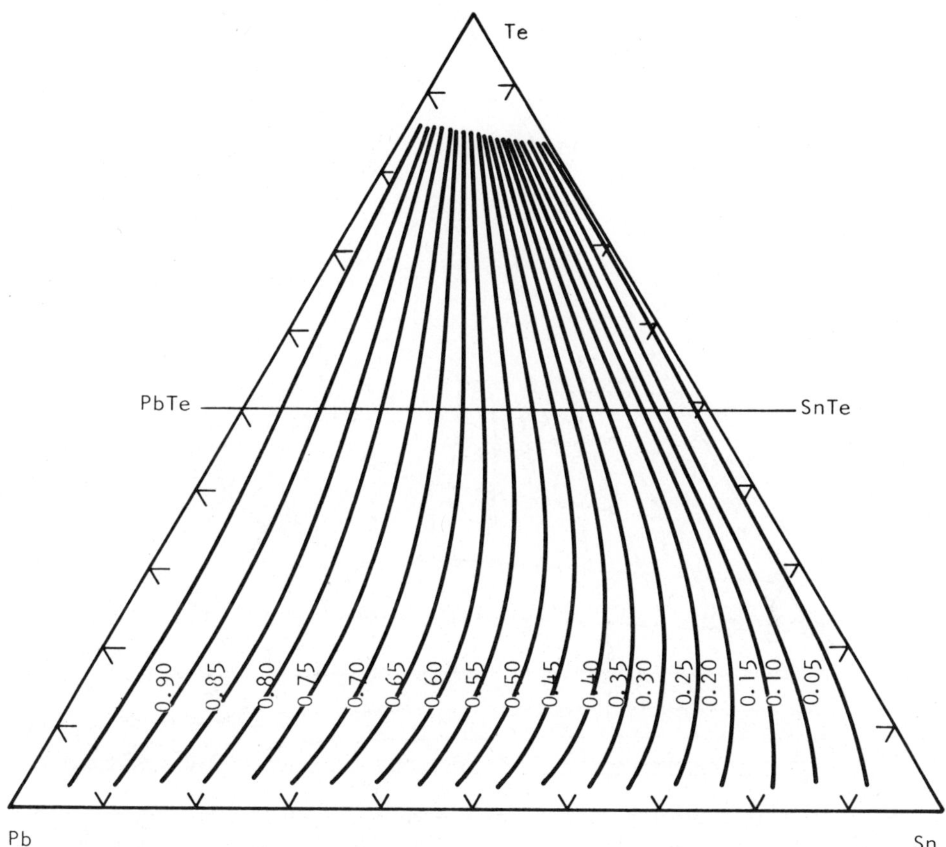

FIG. 3(b). Calculated isosolid composition lines of Pb-Sn-Te. Each line represents the loci of liquid compositions in equilibrium with a given solid composition. Each line is labeled with the mole fraction PbTe in the solid. [From A. Laugier, J. Cadoz, M. Faure, and M. Moulin, *J. Cryst. Growth*, *21*, 235 (1974); reprinted with permission.]

Representative data are shown in Fig. 5. In the single-phase region, a_0, the lattice parameter, varies rapidly with a change in y, while outside the single-phase region changes in y have little effect on a_0. Brebrick reported that for x = 0.80, 0.90, or 1, the metal-saturated limit at 400°C was y = 0.5000 ± 0.0002. However, the tellurium-saturated limit at 356°C was y = 0.5070 for x = 1, y = 0.5056 for x = 0.90, and y = 0.5038 for x = 0.80, a reversal of the trend at higher temperatures near the melting point.

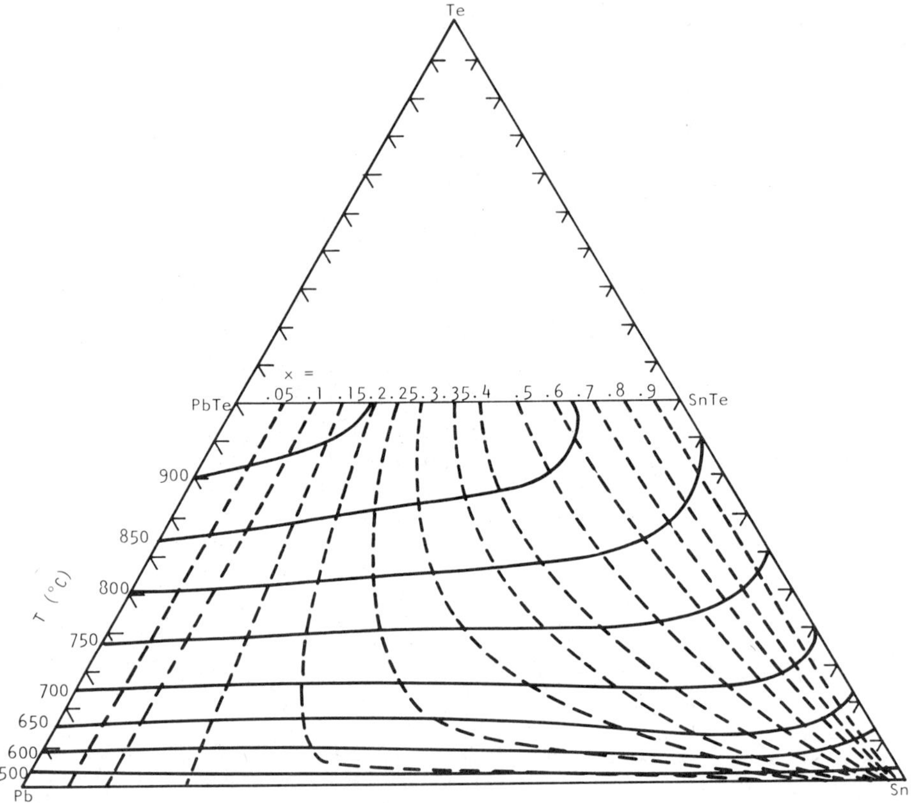

FIG. 4. Calculated liquidus and solidus lines for the ternary system Pb-Sn-Te. The solid lines are isotemperature lines while the dashed lines are isocomposition solidus lines. [From J. S. Harris, J. F. Longo, E. R. Gertner, and J. E. Clark, *J. Cryst. Growth*, *28*, 334 (1975); reprinted with permission.]

Bis and Dixon [12] studied the relationship between a_0, x, and stoichiometry in $Pb_{1-x}Sn_xTe$. They showed that Vegard's law is obeyed only for small deviations from stoichiometry. They assumed that the effect of vacancies on a_0 was the same for all values of x and that the carrier-vacancy ratio remained constant. Data for two carrier concentrations are shown in Fig. 6(a). For carrier concentrations $>10^{20}$ cm^{-3}, Vegard's law is no longer obeyed [10,12-14]. This is illustrated in Fig. 6(b) from the study by Bis and Dixon [12] who used the experimental data of others [10,13,14] to show deviations

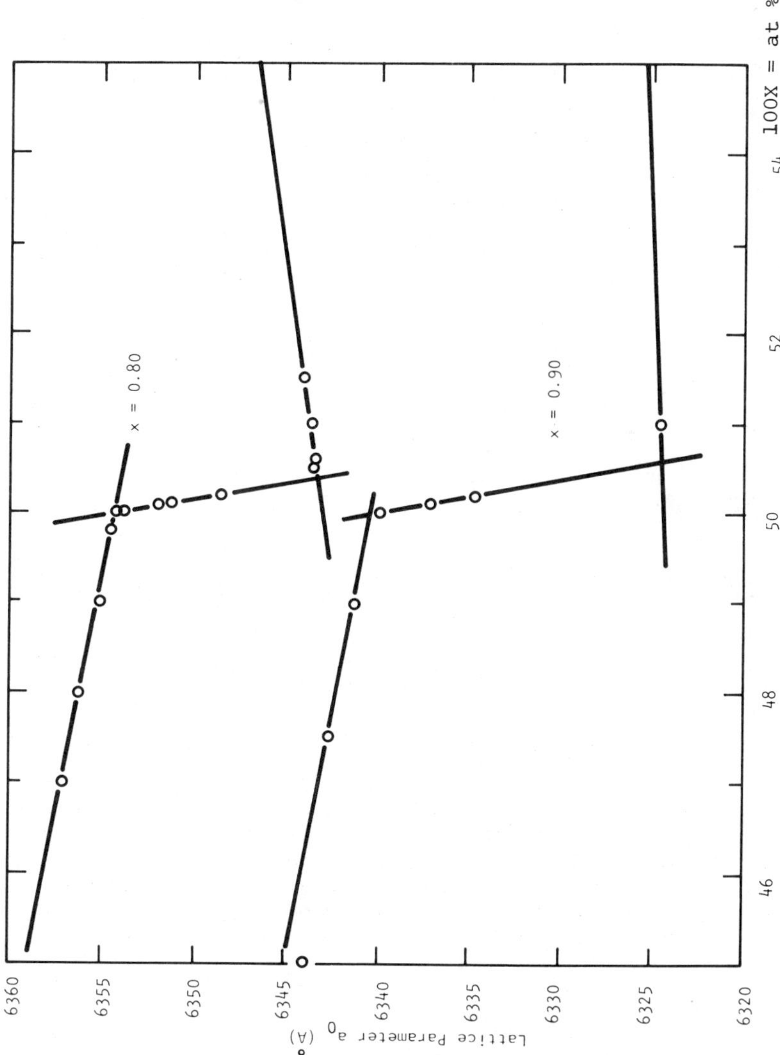

FIG. 5.   Lattice parameters of $(Pb_{1-x}Sn_x)_{1-y}Te_y$ as a function of y for values of x = 0.80 and x = 0.90. The rocksalt-telluride solid solution occurs along the middle line segments for each value of x. (Reprinted with permission from *Journal of Physical Chemistry*, Vol. 32, R. F. Brebrick, "Composition Stability Limits for the Rocksalt-Structure Phase $(Pb_{1-y}Sn_y)_{1-x}Te_x$ from Lattice Parameter Measurements," p. 556, Copyright 1971, Pergamon Press, Ltd.)

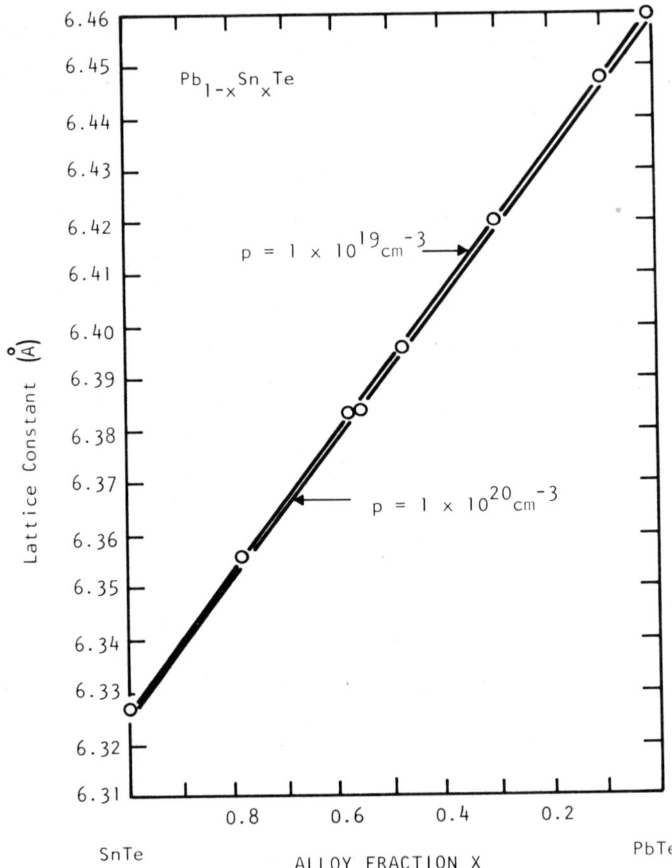

FIG. 6(a). Vegard's law as applied to Pb$_{1-x}$Sn$_x$Te alloys at two carrier concentration each corresponding to a fixed deviation from stoichiometry. The data points represent experimental determinations of the lattice constant and alloy fraction for alloys having carrier concentrations [p = 1/(R$_{He}$)$_{300K}$] ranging from 1.2 × 10^{18} to 7.5 × 10^{19} cm^{-3}. (From Ref. 12; reprinted with permission.)

from Vegard's law. It should be noted that these deviations are more pronounced with increasing SnTe concentration. Many of the inconsistencies and deviations from Vegard's law were due to the high carrier concentrations caused by the alloy compositions and by the growth-anneal methods used.

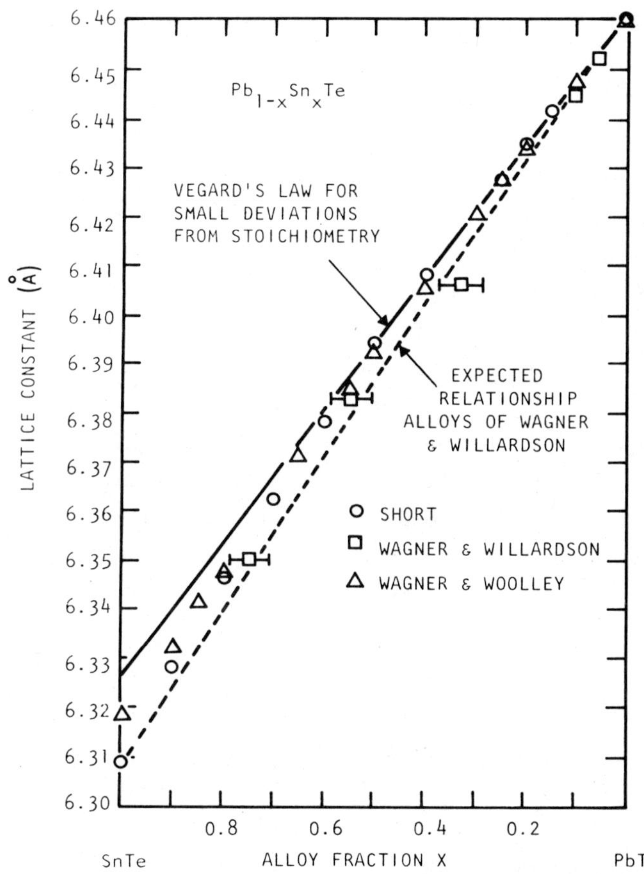

FIG. 6(b). Variation of lattice constant with composition arising
from deviations from stoichiometry. The expected variation represented by the dashed line was calculated and is in agreement with the
experimental data of Wagner and Willardson [10]. The data of Wagner
and Woolley [13] and of Short [14] show similar variations. The
sample preparation methods employed by the three groups were not
identical. (From Ref. 12; reprinted with permission.)

III. MATERIAL PREPARATION AND BULK CRYSTAL GROWTH

Many techniques have been used to grow crystals of (Pb,Sn)Te, but
large single crystals of high perfection are difficult to prepare.
The starting elements are easily purified and compounded to give high
chemical purity. Melting points are not excessive, $\leq 920°C$, and the

solid compound does not undergo any phase transformations between
room temperature and melting. However, the combination of the liquidus-
solidus separation, the low material strength, and a susceptibility
to inclusions, precipitates, grain growth, and other defects generally
leads to low-quality crystals of limited size.

A number of workers have used the Bridgman technique, or modifications of it, for growth of (Pb,Sn)Te crystals. Calawa et al. [4] used
a carbon-coated fused silica sealed ampoule. They compounded from a
stoichiometric mixture and used a tapered point on the ampoule to
facilitate single-crystal growth. The ampoule was lowered through a
temperature gradient of $\sim 5°C$ cm^{-1} at 1 cm day^{-1}; the ingot varied in
composition along its length, was polycrystalline, and contained precipitated Te. Dionne and Woolley [15] used a modified Bridgman technique and found small-angle grain boundaries in about half of a typical
crystal. Butler and Harman [16] observed metallic inclusions and cellular substructure in (Pb,Sn)Te crystals grown by the Bridgman method.
The defects, revealed by an electrolytic etch technique [17], were
perhaps the result of constitutional supercooling at the growth interface. Effects of constitutional supercooling can be lessened by increasing the thermal gradient or perhaps by changing the melt composition, but the overall problems inherent in the Bridgman method seem
to prevent preparation of good crystals.

Wagner and Willardson [18] grew single crystals of (Pb,Sn)Te
from nonstoichiometric metal-rich melts by the Czochralski technique
with liquid encapsulation under boric oxide. They found that small-
angle grain boundaries could be reduced or eliminated by necking the
seed diameter to 1 or 2 mm for several mm. Pull rates of 1 to 3 mm
hr^{-1} were used to grow single crystals from melts containing as little
as 30 at% Te. In runs with the lower amounts of Te, the growth rate
had to be reduced to obtain single-crystal material. The crystals
grown were 0.5 to 1 cm diameter and 2 to 3 cm long and weighed 10 to
20 g; attempts to grow larger-diameter crystals usually resulted in
polycrystalline material. Only p-type material with 10^{18} cm^{-3} carriers
could be produced by this method.

Kimura [19] used a high-temperature recrystallization method to grow single crystals from a polycrystalline ingot. A temperature and composition were selected so that a small fraction of the ingot was liquid. A small thermal gradient, 2 to 3°C cm^{-1}, was present and the liquid was at the colder end where it could serve as a source or sink for diffusion during the solid state crystal growth. The single crystal, up to 1 cm in dimension, occupied about half the ingot and had a constant Pb/Sn ratio. No metallic inclusions or cellular substructure were found in the single-crystal portion of the ingots. Single crystals of $Pb_{1-x}Sn_xTe$ were grown with x = 0.178 to 0.241 with 2.8 to 3.0×10^{19} cm^{-3} p-type carriers.

To eliminate the problems of growing (Pb,Sn)Te from the melt, various vapor growth methods have been used to produce bulk crystals. PbTe vaporizes congruently, i.e., the vapor is PbTe molecules. SnTe vaporizes noncongruently, but the dissociation is very slight and in a closed system would lead to a slight excess of Te in the vapor [20-22]. The vapor pressures of PbTe and of SnTe at 1100 K are 2.13 and 2.66 torr, respectively [21]. Thus, a crystal grown by vapor transport from a (Pb,Sn)Te charge initially contains slightly more SnTe than the charge. In runs where SnTe samples were completely sublimed, small amounts of metallic Sn (about 1% of the initial sample) were left behind [23].

Sealed-tube techniques have been used for crystal growth by vapor transport techniques [24], but the initial work produced only small crystals. Butler and Harman [25] used self-seeding in a closed tube to grow small platelets of (Pb,Sn)Te, a few mm on an edge. They used a temperature difference of ∿2.5°C between the charge and the crystal growth region; the charge was not completely transported. The growth tube was 13 cm in length and 15 mm ID and was tapered in the crystal growth region. The growth tube was held vertically in the furnace and the growth charge was held in an inner quartz tube. After crystal growth they annealed the crystal platelets in the unopened tube by lowering the temperature stepwise over a period of several days. By such a method, with metal-saturated charges, they were able to produce

n-type layers a few μm thick on the p-type crystals which are usually
formed during growth.

Ohtsuki et al. [26] grew single crystals of (Pb,Sn)Te about 1
cm^3, uniform in composition, in a sealed tube by vapor transport onto
a flat fused silica plate. The (Pb,Sn)Te alloy was sealed at 10^{-6}
torr in the tube. The plate region was first heated alone at 820°C
for 2 hr, then the entire tube was inserted into the furnace. The
charge was at 800 to 820°C with the plate 3 to 6°C cooler. Nucleation was random, but occasionally large single crystals resulted.
The crystals contained $\sim 1 \times 10^5$ cm^{-2} dislocations and had hole mobilities several times greater than crystals grown by the Czochralski
technique.

Pandey [27] used a vapor transport technique to grow oriented
bulk single crystals of (Pb,Sn)Te on BaF_2 substrates. The source
material was held at 850°C with the BaF_2 substrate 4 to 10°C lower.
Both funnel-shaped and cylindrical deposits were grown; the funnel-
shaped deposits were more likely to contain large single crystals.
By carefully adjusting temperature gradient between source and substrate and by providing a good heat sink over the entire area of the
substrate, it was sometimes possible to grow a defect-free cylindrical
single crystal. Good growth was dependent on careful handling of the
BaF_2 substrates, which break easily because of thermal stress. No
properties of the crystals were given.

Single crystals of (Pb,Sn)Te have been grown by [28,29] using
an ingot nucleation technique from a source containing 0.01% excess
Te. The Pb, Sn and Te, 99.9999% pure, were melted together for 16 hr
at 1000°C in an evacuated and sealed fused silica tube without melt
agitation. After quenching in water, the first-to-freeze end of the
ingot was discarded; in addition, about 20% of the ingot was sublimed
off in vacuum. For growth, the source ingot was sealed off at 10^{-6}
torr in a 15-mm ID 15-cm long fused silica ampoule. Growth temperatures of 805 and 835°C were used with growth periods of 12 to 14 days;
the horizontal temperature difference along the ampoule was 2°C.
Crystals up to 1 cm on an edge were grown on the quartz walls and on

the source material. The crystals grown on the source material were at least an order of magnitude lower in dislocation density (1×10^3 cm^{-2}) than those grown on the walls. Compositions of $0.06 < x < 0.08$ were grown and had $\sim 10^{19}$ cm^{-3} p-type carriers.

Harmon and McVittie [30] also used a horizontal, unseeded vapor growth (HUVG) method to obtain crystals with facets as large as 2.5 cm^2 surface and good crystalline perfection. PbTe and SnTe were prepared from 99.9999% elements and were zone-refined under B_2O_3. The desired ratio of compounds plus small amounts of excess metal or tellurium if desired were reacted in an evacuated ampoule for 4 hr at 50°C above the liquidus temperature, then water-quenched. This was followed by isothermal annealing in the reaction tube at 800°C for 5 days, then air-quenched. The source material was crushed with an agate mortar and pestle to ~ 1.5 mm. A charge of ~ 40 g was uniformly distributed inside a 2.5 cm ID $\times$ 15-cm evacuated sealed fused silica ampoule and heated in a horizontal furnace. The horizontal and vertical temperature gradients measured in the absence of the ampoule were 0.1 to 0.2°C cm^{-1} and 1.0 to 1.4°C cm^{-1}, respectively. In the vertical direction the highest temperature occurred at the bottom of the ampoule. Growth temperatures of 800 to 815°C were used for a typical growth period of 7 days. The crystals grew on the surface of the powdered charge. At the end of the run, the ampoule was cooled in the furnace by turning off the power; this gave a lower dislocation density than air-quenching.

The above preparation and cleanup of the charge before crystal growth apparently is necessary to obtain good, large crystals; use of as-received Pb, Sn, and Te gave little or no crystal growth. Excess tellurium caused lowered growth rates, and metallic inclusions were found in the crystals if source material exceeded 50.1 at% metal. Otherwise, crystals grown by the HUVG technique were free of cracks and low-angle grain boundaries on the facets, though some voids were present 0.1 to 0.3 cm below the facet. Facet areas up to 1 mm^2 were free of dislocations and the workers found concentrations as low as 10^3 cm^{-2} over the entire facet. A major advantage of this method is

that crystals grow on the source material rather than in contact with
the ampoule wall, so crystal strain and damage can be greatly reduced.

Large high-quality crystals of (Pb,Sn)Te, 2.5 cm diameter × 5 cm
weighing up to 250 g, have been grown by Parker et al. [31,32]. Crystals were grown by self-transport in a sealed system onto oriented
seeds in an apparatus shown in Fig. 7(a). Grown crystals are shown
in Fig. 7(b). Starting elements, 99.9999% pure, were further purified
by zone-refining under hydrogen. The charge was sealed in an evacuated
fused silica ampoule and reacted in a rocking furnace for 16 hr at

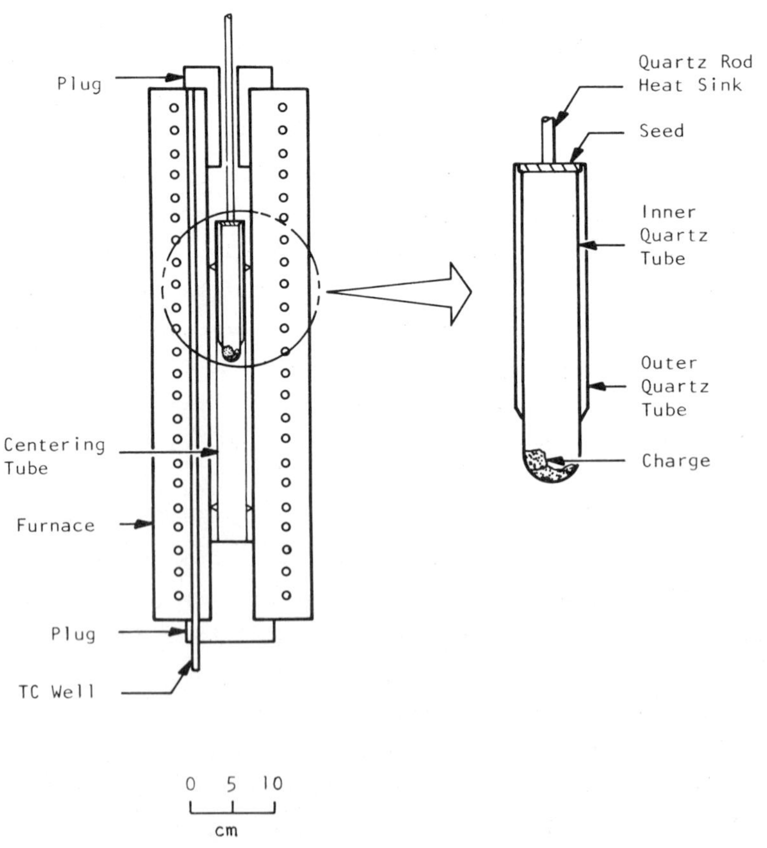

FIG. 7(a). Sketch of apparatus used to grow large crystals of Pb_{1-x}
$Sn_x Te$ by vapor transport. (From Refs. 31 and 32; reprinted with
permission.

1000°C, then oil-quenched. The ingot was removed to another sealed,
evacuated fused silica ampoule and heat-treated at 800°C for ∿15 days.
During this treatment, a portion of the ampoule was at ∿750°C and
∿25% of the charge transported to the cooler end. We felt that this
treatment homogenized the ingot by reducing nonstoichiometry and by
ensuring complete reaction of all elements.

The residue from the above heat treatment was used as the charge
for growth on a (111) polished single crystal of (Pb,Sn)Te. A constant-
diameter opening from seed to growing crystal produced crystals with
less strain and fewer low-angle grain boundaries than did a tapered
opening. The sealed evacuated tube was put in a vertical furnace with
the charge at 825 to 850°C and the seed 1 to 5°C cooler. Growth rates
were 5 to 8 g day^{-1} for up to 5 weeks; about 75% of the charge was

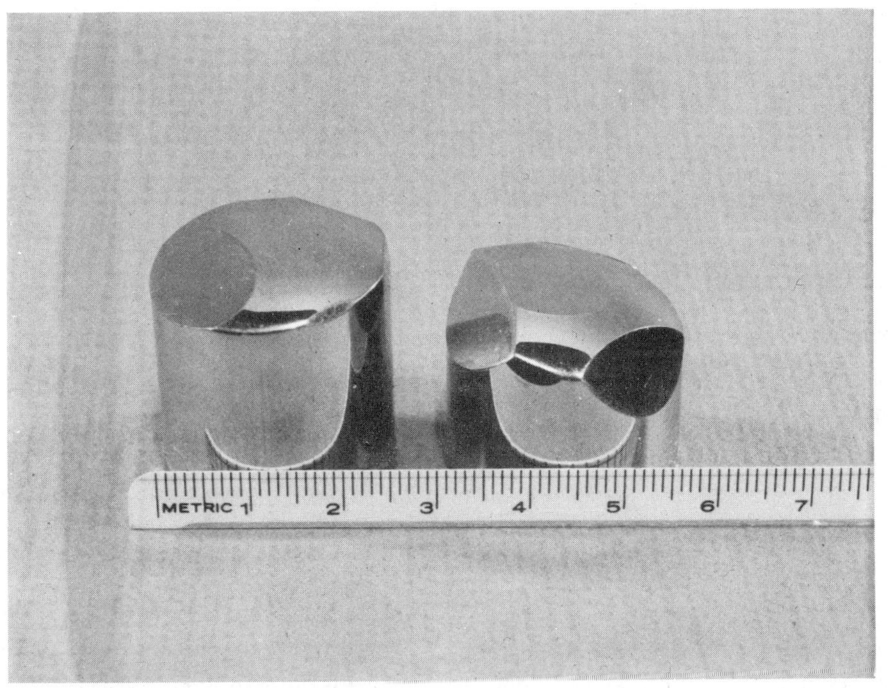

FIG. 7(b). (Pb,Sn)Te single crystals grown by vapor transport. (From
Ref. 32; reprinted with permission.)

transported. At the end of the run, the furnace was cooled to room
temperature at $10°C$ hr^{-1}. The resulting material was free of voids
and inclusions with no small-angle grain boundaries or strain and
with 10^2 to 10^3 cm^{-2} dislocations. By suspending the seed on wall
indentations with no other contact and by using a growth gradient
(source to seed) of $1°C$, crystals were grown with 10 to 10^2 cm^{-2}
dislocations. Wall contact or constriction seems to introduce strain,
perhaps by the difference in thermal coefficients of expansion. For
optimum crystal perfection, it seemed necessary to start with a charge
with 0.01 at% excess Te. For a crystal of $Pb_{0.79}Sn_{0.21}Te$, p = 3 ×
10^{19} cm^{-3}. Linear growth rates of 0.5 to 1.0 mm day^{-1}, averaged from
seed to final growth interface, were possible.

Tamari and Shtrikman [33] developed a similar growth method but
used self-nucleation on a fused silica plate as a source of seeds.
They report single crystals of $Pb_{(1-x)}Sn_xTe$, $0 \leq x \leq 0.25$, up to 100
g and 40 mm diameter. By using a slight supercooling at the start of
growth, they encouraged a relatively few nuclei or a single crystal
to start; additional growth could be accelerated by additional under-
cooling. Crystals contained a few holes and voids and had about 10^4
cm^{-2} dislocations.

IV. EPITAXIAL FILM GROWTH

For many applications, (Pb,Sn)Te crystals of only a few micrometers
thickness are needed; it is then more practical to deposit thin layers
than to lap and polish bulk material to the desired thickness. In
addition, epitaxial growth has the potential to produce either n- or
p-type material with a range of carrier concentrations and to deposit
layers on insulating substrates, such as BaF_2. Both vapor transport
and liquid-phase techniques have been used for epitaxial deposition
of (Pb,Sn)Te.

Bylander [34] vacuum-evaporated films of (Pb,Sn)Te onto KCl substrates using a previously compounded source alloy. Few details were
given on the experimental procedures and the physical properties of
these films but source and films were approximately the same composition.

For $Pb_{0.75}Sn_{0.25}Te$, the hole concentration was 3 to 5 × 10^{18} cm^{-3} with mobilities up to 8800 cm^2 V^{-1} sec^{-1} at 77 K.   Holloway et al. [35] used a two-source (PbTe and SnTe) evaporation technique to deposit (Pb,Sn)Te on BaF_2 substrates.   The double Knudsen cells were heated to 700°C and layer composition was selected by appropriate choice of orifice sizes.   Films with the best properties were obtained with the substrate at 325°C during deposition; however, even these films contained a mixture of (111) and (100) orientations.   The films were not uniform in composition from bottom to top.   Mathur [36] used molecular beams of the three elements to obtain (Pb,Sn)Te films of uniform composition.   Bis et al. [37] used evaporation from a single-alloy source to obtain (Pb,Sn)Te films up to 328 μm thick on NaCl substrates.   A heated fused silica tube connecting the source and substrate helped to establish quasi-equilibrium conditions so that good-quality films were obtained.   The substrate was at 325°C and the source and tube were at 530°C.   The carrier concentration in these films was uniform to 10% but it was not determined if the films contained a mosaic structure.   Lowney and Cammack [38] used a similar apparatus but used cleaved KCl substrates.   Better-quality films were obtained by improving the adherence of films to the substrate; this was done by depositing a 1-μm thick film at 280°C, then raising the substrate to 400°C for further deposition.   Kasai et al. [39] also used a modified version of the Bis [37] hot wall technique to deposit from a source of $Pb_{1-x}Sn_{x+\delta}Te_{1-\delta}$ onto polish-etched KCl substrates. Both n- and p-type layers were obtained from either stoichiometric or metal-rich sources.   Carrier concentrations of as-grown layers with $0 \leq x \leq 0.2$ were in the range of 10^{16} to 10^{17} cm^{-3} and mobilities were about 10^4 cm^2 V^{-1} sec^{-1} at 77 K.   Both n- and p-type layers of high quality were obtained from metal-rich sources with $\delta = 0.01$ by controlling the Te vapor pressure from a separate reservoir.

Corsi [40] deposited (Pb,Sn)Te layers in an RF multicathode sputtering system capable of simultaneous or sequential sputtering at different rates from three different targets.   Epitaxial films were obtained on Ge, NaCl, and BaF_2 substrates.   It was possible to obtain n- or p-type layers by controlling the quantity of excess metal or Te, respectively. Rotation of the substrates affected the deposition rates and consequently the electrical properties, particularly on BaF_2 substrates, but no quantitative measurements were made. Mobilities higher than 10^4 cm^2 V^{-1} sec and carrier concentrations lower than 10^{17} cm^{-3} at 77 K were obtained in p-type samples deposited on NaCl substrates.

Rolls et al. [41] produced single-crystal epitaxial layers of $Pb_{1-x}Sn_xTe$, 0.15 < x < 0.19, on PbTe substrates by an open-tube vapor-phase transport system. Deposition substrates oriented (100) gave the best crystalline properties. Films up to 100 μm thick were obtained. The $Pb_{1-x}Sn_xTe$ charge was compounded from the elements in an evacuated silica ampoule, air-quenched, and annealed at 600°C for 1 week. A fused silica growth chamber was used with a source temperature of 725°C and hydrogen carrier gas flowing at 300 to 1000 ml min^{-1}. Growth rate depended on the position of the substrate in the temperature gradient. With a (100) PbTe substrate at 640°C, the growth rate was 10 μm hr^{-1}. Deposition on (111) or (110) PbTe gave poor-quality films. Bellavance and Johnson [42] used a similar method to grow (Pb,Sn)Te on high-quality (100) (Pb,Sn)Te substrates, using care in the purity of the source material. High-quality films were obtained only when the source was stoichiometric; excess metal or Te caused inclusions and voids in the films. With a source of $Pb_{0.79}Sn_{0.21}Te$, growth was obtained at substrate temperatures of 425 and 550°C with p-type carriers of 3×10^{16} and 3×10^{17} cm^{-3}, respectively. Growth rates were 3 to 4 μm hr^{-1}. N-type material could not be produced even though the deposition temperature was below the point where the equilibrium carrier concentration should have been on the n-type side of stoichiometry (see Fig. 2). However, n-type films could be obtained by using a dopant such as Sb or by reducing the Sn content (lower values of x). Parker [43] used a closed-tube vapor transport method to deposit (Pb,Sn)Te epitaxial films on (Pb,Sn)Te substrates with results similar to those reported by the open-tube method [42]. The films obtained on the (Pb,Sn)Te substrates contained fewer defects than films on other substrates such as BaF_2 or NaCl, probably because of a better match of thermal coefficients of expansion. The open-tube vapor transport technique of Bellavance and Johnson [42] could be modified for a high production rate of (Pb,Sn)Te films with low carrier concentration.

One problem with vapor growth pointed out by Freik et al. [44] is the change in composition with film thickness. Various workers have noted these changes in composition and have attempted to limit the change by growth of very thin films or by use of very large charges. The observed changes in composition are related to the difference in vapor pressure of SnTe and PbTe [20,21] or nonstoichiometry in the source.

Longo et al. [45,46] deposited epitaxial films of (Pb,Sn)Te on (Pb,Sn)Te substrates using liquid-phase epitaxy from a Pb-Sn solution. High-quality films with p-type carrier concentrations $\leq 10^{16}$ cm^{-3} in $Pb_{0.82}Sn_{0.18}Te$ were obtained. The morphology of the epitaxial films depended on growth conditions at the solid-liquid interface. Thermal gradients of 10 to 60°C cm^{-1} were used to produce good crystalline quality. The growth technique included tilting the apparatus to bring the melt to the substrate and decreasing the temperature at a rate of 0.12 to 0.6°C min^{-1}. Good crystalline morphology occurred only on substrates which were less than 0.1° off (100). Growth on (110) or (111) planes gave very rough surfaces and metallic inclusions.

Tomasetta and Fonstad [47] grew epitaxial films of (Pb,Sn)Te on PbTe substrates from a solution of PbTe in Pb-Sn by a conventional horizontal sliding apparatus technique. In growth, a melt saturated with PbTe and contained in a graphite slider was moved to cover the substrate and then the furnace temperature was lowered. After a temperature drop, the melt was removed by moving the slider. Temperature gradients were not intentionally introduced. Smooth shiny epitaxial layers, free of inclusions, were grown by this method. These workers grew a 15 μm film of $Pb_{0.9}Sn_{0.1}Te$ with a temperature decrease from 600 to 590°C; the film was n-type with 1×10^{17} cm^{-3} carriers. Both n- and p-type films were grown depending on the growth temperature and melt composition. In a similar apparatus, Groves [48] found that minimizing the temperature gradient across the liquid-solid interface

improved surface crystal perfection in the film. Segregation coefficients of Pb and Sn in the sliding apparatus [47,48] were different
than those in the tilting apparatus [46]. Recently, Astles et al.
[49] grew epitaxial films with a sliding boat technique at temperatures as low as 400°C and hole concentrations as low as 5×10^{16} cm^{-3}
at 77 K. High-spatial-resolution thermoelectric power measurements
showed carrier concentration uniformity both laterally across the
slice and through the deposit.

Ickert et al. [50] concluded that gas-phase epitaxy was better
for films of high crystalline perfection, though melt growth methods
were simpler in execution.

V. CRYSTAL DEFECTS AND EXTRINSIC CARRIER CONCENTRATION

(Pb,Sn)Te crystals or epitaxial films may contain the following defects: lattice point defects, dislocations, inclusions, precipitates,
low-angle grain boundaries, holes, linear voids, and crystalline
strain. All of these defects can be minimized by judicious selection
of growth conditions or by careful handling techniques. Lattice
point defects and chemical impurities can give material with high
carrier concentrations; the excess carriers can be controlled by annealing or, to a limited extent, by counterdoping.

Harman [51] zone-refined PbTe and SnTe under B_2O_3 to reduce the
amount of chemical impurities. Parker et al. [31] zone-refined the
elements under flowing H_2 before compounding; typical results after
50 zone passes are shown in Table 1. After compounding [31,51],
the (Pb,Sn)Te contained about the same concentration of impurities
as expected from summing the impurities in the starting elements or
compounds.

In all cases, a major defect in the bulk crystals or films is
point defects, i.e., Pb, Sn, and Te vacancies and/or interstitial
atoms. The concentration of these point defects, in the as-grown
materials, can be predicted from the phase diagrams. For example,
from Fig. 2(b) it can be seen that growth of $Pb_{1-x}Sn_xTe$ compounds
within a few degrees of the liquidus temperature will produce p-type

TABLE 1

Mass Spectrographic Analysis of
Zone-refined Elements (ppm Atomic)

Impurities[a]	Pb	Sn	Te
Cd	0.8	<1	1
Fe	0.4	<2	0.3
K	0.5	<2	1
S	--	0.3	--
Si	0.2	--	0.4
Al	0.2	0.1	0.3
Na	0.3	0.5	0.4
O	4	3	3
N	0.2	--	1
C	1	3	6
Mg	--	<0.9	0.2

[a]These were the only impurities detected.
Source: Unpublished data of S. G. Parker.

carriers $\geq 10^{18}$ cm^{-3}. Even for small values of x, the crossover temperature* is less than 750°C. The crossover temperature varies with
doping level; thus Harman [51] reports ∿635°C as the value for
$Pb_{0.8}Sn_{0.2}Te$ and we have measured a value of ∿600°C. Most melt or
vapor growth techniques are used at temperatures near the maximum
liquidus temperature so high carrier concentration p-type material
usually results.

Butler and Harman [16] annealed crystals in their growth tube
to reduce the carrier concentration. However, the isothermal annealing described by Calawa [4] is the most widely used method to control
carrier concentration in (Pb,Sn)Te. A crystal of arbitrary initial
defect concentration is sealed in an evacuated ampoule with a (Pb,Sn)Te
powder of the same Pb/Sn ratio as the crystal but containing excess

----

*Crossover temperature is that temperature where solid (Pb,Sn)Te, in
equilibrium with a vapor phase, is on the stoichiometric line.

metal or tellurium. Isothermal annealing at a constant temperature
will bring the crystal to the metal-rich or tellurium-rich solidus
boundary of its phase field at that temperature. The data of Fig.
2(b) were acquired in this way.

In an extensive study of annealing $Pb_{1-x}Sn_xTe$ by this method,
Hewes et al. [52] showed that several weeks of annealing may be required to reach equilibrium, particularly for samples requiring low
temperatures. The annealing time can be reduced by annealing at different temperatures as equilibrium is approached. For example, to
reduce p-type carrier concentration from 3×10^{19} to 5×10^{17} cm^{-3}
in $Pb_{0.79}Sn_{0.21}Te$, the material was annealed at 750°C for 2 days,
650°C for 4 days, and 600°C for 7 days [5].

One advantage of both vapor and liquid phase epitaxial [53]
growth is that low growth temperatures may be selected producing low
carrier concentrations with little or no annealing. If annealing is
necessary, then the same techniques may be used as for bulk single
crystals. Another way to reduce the carrier concentration is to in-
corporate dopants in the material during growth. Tomasetta and Fonstad
[54] grew (Pb,Sn)Te films doped with Tl or In using liquid-phase epitaxy. Thallium is a p-type dopant and indium is n-type. They showed
that the doped layers could be annealed under conditions that would
convert undoped material from n- to p-type, or vice versa, without
converting the doped layers. Annealing periods of 4-7 days were used.

Antcliffe and Wrobel [55] used diffusion of Cd or Zn into vapor-
grown samples of (Pb,Sn)Te to obtain n-type material with carrier concentrations of 1.3×10^{17} cm^{-3} and mobilities up to 5×10^5 $cm^2 V^{-1}$
sec^{-1} at 4.2 K. Linden [56] incorporated Cd or Zn into (Pb,Sn)Te crystals during vapor growth at 700 to 800°C. Higher Cd or Zn pressure
gave n-type crystals with carrier concentrations as low as 4.4×10^{16}
cm^{-3}. However, the results were irreproducible, so this method may
be of limited value in obtaining low carrier concentrations.

In the growth of large single crystals by vapor transport [32],
holes or voids were related to the source composition and to growth
conditions. The holes were caused mainly by excess of unreacted Pb

and/or Sn in the charge. Rapid growth also caused formation of holes
and, in some polycrystalline ingots, holes were present at grain
boundaries. The thermal presoak and partial sublimation greatly reduced the number of holes. Chemical analysis showed about 1% more
Pb and Sn in the sublimate than in the residue. Optimum concentration
was found to be 0.01 at% excess Te; additions above 0.1 at% excess
gave Te inclusions in the final crystal.

Parker [32] described another type of defect, a linear void containing internal facets, apparently caused by different growth rates
in different crystal directions. The void was bounded by (100) facets
and was probably caused by a nonuniform heat flow at the growth interface. These voids were minimized by growing from a constant-diameter
opening, by ensuring uniform heat flow from the seed to the heat sink,
and restricting linear crystal growth to <0.7 mm day^{-1}.

(Pb,Sn)Te is a very fragile material and is easily strained or
plastically deformed, even under slight stress [51]. The strain is
best seen with a scanning x-ray topograph or a large-area Laue topograph. If the material is stressed beyond the yield point, dislocations are produced which can be identified by etch techniques [17].
Great care must be taken handling the material; tweezer forces or a
slight drop onto a hard surface introduces dislocations [57]. Rapid
cooling, 100 to 200°C hr^{-1}, in contact with a fused silica tube, gives
10^6 to 10^7 cm^{-2} dislocations; cooling at 10°C hr^{-1} in the same tube
produces $\sim 10^4$ cm^{-2}. A crystal freely suspended and cooled at 200°C
hr^{-1} has 10 to 100 cm^{-2} dislocations [31]. The shape and size of the
growth tube also greatly influences strain developed in bulk crystals.
Material grown in the tapered section of a tube had significantly more
strain than that in the same crystal but in a constant-diameter section of the tube. We also noted [31] that growth in a large-diameter
ampoule, 25 mm ID, gave crystals of little or no strain compared to
severe strains in a crystal from a 10-mm ampoule under otherwise
identical conditions. It was also found that polycrystalline growth
had more strain than single crystals.

VI.   BAND STRUCTURE AND ELECTRICAL PROPERTIES

The band structure of $Pb_{1-x}Sn_xTe$ alloys has been the subject of a
number of studies.   The change in energy gap with temperature and
composition is illustrated in Fig. 8 after Melngailis and Harman [5].
It can be seen that the energy gap close to 0 K is zero at $x \sim 0.35$
and increases on either side of this composition.   In addition, the
temperature coefficient of the energy gap reverses sign at $x = 0.35$;
it is negative on the Pb-rich side and positive on the Sn-rich side.
The data for Fig. 8 were obtained from laser emission [58,59], optical
absorption [34,60], and from tunneling experiments [61].   Recently
Takasaki and Tanaka [62] measured the energy gap using tunneling

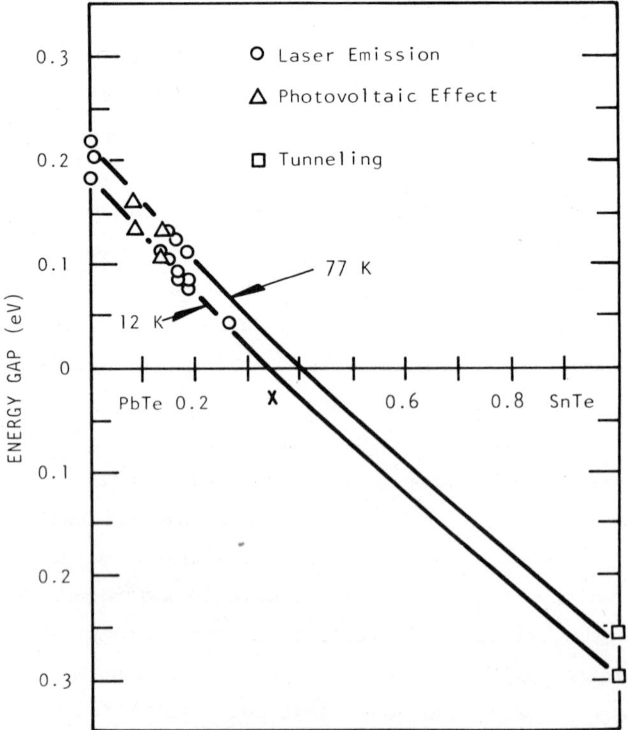

FIG. 8.   Energy gap of $Pb_{1-x}Sn_xTe$ as a function of x.   (From Ref. 5;
reprinted with permission.)

experiments at liquid helium temperature for compositions near the crossing point and confirmed the data of Fig. 8. The composition and temperature dependencies of the energy gap for $Pb_{1-x}Sn_xTe$ may be expressed by

$$E_g \; (eV) = 0.180 - 0.48x + 4 \times 10^{-3} \; T \; (K)$$

Dixon and Bis [63] tested the band-inversion model by measuring the electrical resistivity of $Pb_{1-x}Sn_xTe$ alloys as a function of temperature from 4 to 300 K. The resistivity showed a linear dependence on temperature with a distinct break at the temperature predicted for zero energy gap for the particular composition. The break temperature was independent of carrier concentration and carrier mobility; the linearity was attributed to degenerate lattice scattering. Gurieva et al. [64] obtained similar results which agreed with the band inversion model. They reported that for $x > 0.3$, anomalous effects were seen in the effective mass density of states dependence on composition and temperature; they measured the transverse Nernst-Ettinghausen effect. The anomalies were presumed to be due to the sign change of the temperature coefficient of the energy gap at the crossover composition. From the temperature dependence of the thermoelectric parameters for (Pb,Sn)Te, Freik et al. [65] showed that an inversion of the valence and conduction bands occurred for a change in phase composition. Confirmation of the band inversion was obtained by Walz [66] using absorption edge data. Thus, all evidence points to a band inversion model to explain the band behavior with temperature and compositional changes. A schematic representation of the bands in $Pb_{1-x}Sn_xTe$ as proposed by Dimmock et al. [58] is given in Fig. 9. This model seems to be the most consistent with all experimental results [5,63,67]. The valence and conduction band edges in PbTe occur at the L point in the Brillouin zone with the valence band edge an L_6^+ state and the conduction band edge an L_6^- state [68]. As x increases, the energy gap initially decreases as the L_6^+ and L_6^- states approach each other, goes to zero at some intermediate composition where the two states become degenerate, and then increases with the

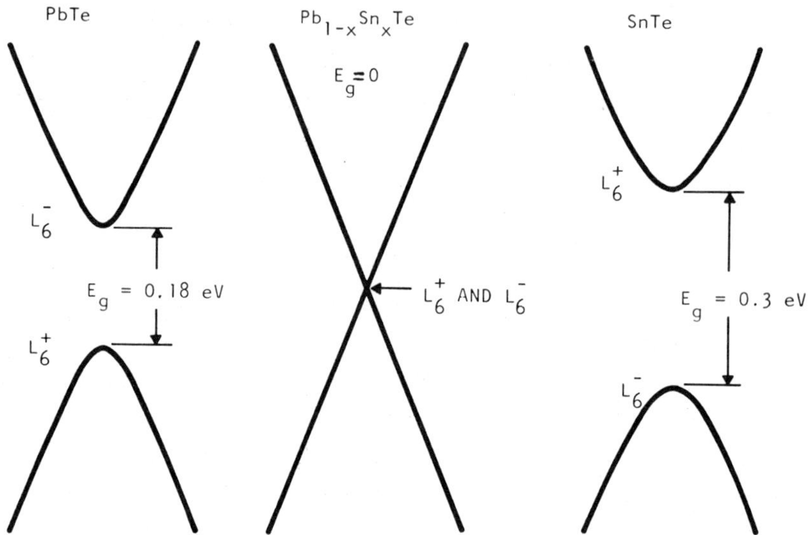

FIG. 9. Schematic representation of the valence and conduction bands at 12 K for PbTe, for the composition at which the energy gap is zero, and for SnTe. (From Ref. 58; reprinted with permission.)

L_6^+ state now forming the conduction band edge and the L_6^- state forming the valence band edge. Because the L_6^+ and L_6^- states each have only a twofold spin degeneracy, their crossover does not result in a semimetal but gives a semiconductor with the valence and conduction bands interchanged. Conklin et al. [68] showed that relativistic corrections are very important in determining the positions of the energy bands; the differences in the relativistic effects [69] in Pb and Sn can be used to explain the variation of the energy gap with composition.

Dimmock [67] developed a $\vec{k}\cdot\vec{p}$ perturbation model for the valence and conduction bands of (Pb,Sn)Te in which the interaction between the extremal valence and conduction bands was treated exactly while the interaction with more distant bands was treated by a second-order perturbation theory. One result was a deviation from the two-band prediction that $g = 2m/m^*$ such that on the Pb-rich side of the crossover $g < 2m/m^*$ and on the Sn-rich side $g > 2m/m^*$, where g refers

to the g factor and m and m* refer to electron mass and effective
electron mass, respectively. The effective electron mass was found
to vary with carrier concentration. Adler et al. [70,71] developed
a $\vec{k}\cdot\vec{p}$ model for the magnetic energy levels of the valance and conduct-
ance band extreme in (Pb,Sn)Te. This model extended the six-band
model of Dimmock to interaction with an arbitrarily directed magnetic
field and included calculation of eigenfunctions, which were used to
account for the Knight shift.

Various other band structures have been proposed to explain the
electrical, optical, and thermal properties of (Pb,Sn)Te; these include such features as multiple valence bands [72-79] and overlapping
valence and conduction bands [80,81]. However, Dixon and Bis [63]
pointed out that these models did not hold for all experimental data.
Breaks of the type observed in the temperature dependence of the resistivity [63] could also result from phase changes, but x-ray studies
on (Pb,Sn)Te have shown only a cubic NaCl structure.

Seddon et al. [82] used ultrasonic measurements on (Pb,Sn)Te to
determine if either band inversion and band crossover effects of phonon
mode-softening effects influence the elastic properties. Table 2
lists the elastic constants and some other physical properties of
PbTe and SnTe and their alloys. When the square of the velocity of
each ultrasonic mode which propagates down the (110) direction in
$Pb_{0.47}Sn_{0.53}Te$ was plotted vs. temperature, there was no evidence of
a change in slope of the elastic constants in the $L_6^+ \longrightarrow L_6^-$ band
crossover region. Thus the elastic constants of $Pb_{1-x}Sn_xTe$ do not
depend markedly upon composition and do not evidence any significant
degree of acoustic phonon mode softening. By contrast, in $Ge_xSn_{1-x}Te$
alloys in the vicinity of the ferroelectric Fcc-Fcr phase transition
temperature, the elastic constants decrease markedly and there is an
attenuation peak. The results also provided information about lattice
dynamics and interatomic binding forces in IV-VI compound alloys.

Korn and Braunstein [83] determined the electric susceptibility
effective mass m*, optical dielectric constant ε_∞, and relaxation life-
time τ, and infrared reflectivity data. These data are presented in

TABLE 2

Elastic Constants of PbTe and SnTe and Their Alloys

	PbTe	$Pb_{0.71}Sn_{0.29}Te$	$Pb_{0.50}Sn_{0.50}Te$	$Pb_{0.47}Sn_{0.53}Te$	$Pb_{0.05}Sn_{0.95}Te$	SnTe
Lattice parameter a_0 (Å)	6.461	6.418	6.387	6.382	6.320	6.304
Density ρ (g cm^{-3})	8.25	7.77	7.14	7.35	6.60	6.51
N/V (atoms cm^{-3}) × 10^{22}	2.966	3.026	3.070	3.075	3.169	3.183
θ_D Debye temperature (K) (calculated from room temperature C_{ij})	159.7	164.7	181.7	180.0	187.1	170.5
Velocities of wave propagated down the [110] direction (cm sec^{-1}) × 10^5						
v (longitudinal)	(2.95)	2.93	4.08	3.29	3.32	3.171
v (shear q//[001])	(1.28)	1.30	1.344	1.38	1.47	1.22
v (shear q//[1$\bar{1}$0])	(2.47)	2.60	3.24	2.66	2.79	2.869
Elastic stiffness constants (dyn cm^{-2}) × 10^{11}						
C_{11}	10.80	10.6	18.10	11.73	11.0	10.93
C_{12}	0.77	0.14	0.31	1.34	0.69	0.21
C_{44}	1.34	1.31	1.30	1.41	1.43	0.969

Property						
Elastic compliance constants $(cm^2\ dyn^{-1}) \times 10^{12}$						
S_{11}	0.935	0.941	0.582	0.873	0.917	0.916
S_{12}	-0.062	-0.012	-0.085	-0.089	-0.054	-0.017
S_{44}	7.46	7.62	7.690	7.69	6.990	10.3
Linear compressibility $(cm^2\ dyn^{-1}) \times 10^{-12}$	0.810	0.917	0.412	0.694	0.808	0.881
Volume compressibility $(cm^2\ dyn) \times 10^{-12}$	2.43	2.75	1.24	2.08	2.43	2.64
Bulk modulus $(dyn\ cm^{-2}) \times 10^{12}$	0.411	0.364	0.810	0.480	0.412	0.378
Young's modulus $(dyn\ cm^{-2}) \times 10^{12}$						
[100] in (001) plane	1.07	1.06	1.718	1.14	1.09	1.09
[110] in (001) plane	0.436	0.423	0.461	0.462	0.459	0.330
[111] in (110) plane	0.363	0.352	0.370	0.402	0.385	0.281
Hole concentration (cm^{-3})	3×10^{18}	5×10^{19}	6.8×10^{19}	6.6×10^{19}	6.7×10^{19}	$4.5 \times\times 10^{20}$

Source: Ref. 82; reprinted by permission of Chapman and Hall Ltd.

TABLE 3

Values of m^*, ε_∞, and τ for $Pb_{1-x}Sn_xTe$

Sample	p-Type Hall carrier[a] concentration (cm^{-3})	77K			300K	
		ε_∞	m^* (Electron mass)	τ (10^{-15} sec)	ε_∞	τ (10^{-15} sec)
PbTe	8×10^{18}	35.4	0.102	29.9	34.3	19.6
$Pb_{0.83}Sn_{0.17}Te$	1.0×10^{20}	35.1	0.423	37.2	35.7	20.7
$Pb_{0.63}Sn_{0.37}Te$	2.3×10^{20}	34.7	0.453	13.6	32.3	9.5
$Pb_{0.50}Sn_{0.50}Te$	4.5×10^{20}	35.3	0.385	10.4	39.3	11.0
$Pb_{0.25}Sn_{0.75}Te$	6.5×10^{20}	37.8	0.316	7.29	35.5	5.06
$Pb_{0.12}Sn_{0.88}Te$	7.5×10^{20}	39.0	0.279	9.45	37.1	5.69
SnTe	8×10^{20}	35.9	0.256	8.61	38.8	5.89

[a]Carrier concentration data calculated from Hall measurements. These data were used to calculate m^*.
Source: Ref. 83; reprinted with permission.

Table 3. The data [84-86] were fit to a single-band Drude free carrier model by a least-squares method. Note that the dielectric constant is relatively high and varies little with temperature and composition while the effective mass and relaxation time vary with composition. The dependence of m^* on composition indicates band "flut-
ing" as predicted from band calculations [87-89].

Effective mass has also been determined by nuclear magnetic
resonance and by infrared cyclotron resonance [90,91]. Optical dielectric constant measurements of $Pb_{1-x}Sn_xTe$ were extended to 6 K for
values of x from 0 to 0.4 [92]; these authors found ε_∞ increased to
60 at 6 K for $Pb_{0.61}Sn_{0.39}Te$. Takano et al. [93] measured the static
dielectric constant of $Pb_{1-x}Sn_xTe$ for x of 0 to 0.29 at 1.6 and 4.2 K
using microwave magnetoplasma effects at 50 GHz. The dielectric constant increased to $\sim 10^4$ at 1.6 K for x = 0.29; they saw no indication
of a ferroelectric phase transition as predicted by Bate et al. [94].

Bate [95] calculated the intrinsic carrier concentration in
(Pb,Sn)Te alloys by assuming that the similarity of the density of
states in the valence band maintains both the electrons and holes non-
degenerate in intrinsic material at all temperatures of interest.
Under these conditions, the intrinsic carrier concentration is given
by

$$n_i = 4.9 \times 10^{15} \left(\frac{m_{de} m_{dh}}{m_0^2} \right)^{3/4} T^{3/2} e^{-E_g(1/2)kT}$$

where E_g is the energy gap and m_{de}/m_0 and m_{dh}/m_0 are the density of
states effective mass ratios for electrons and holes, respectively.
These mass ratios were calculated as a function of energy gap using
a six-band $\vec{k}\cdot\vec{p}$ model. Figure 10 shows the calculated carrier concentration as a function of bandgap at various temperatures. Because
the energy gaps in $Pb_{1-x}Sn_xTe$ are temperature-dependent, the curves
do not refer to a specific alloy composition.

Machonis and Cadoff [96] determined α, the Seebeck coefficient,
for $Pb_{1-x}Sn_xTe$ alloys up to x = 0.4; their data are shown in Fig. 11.

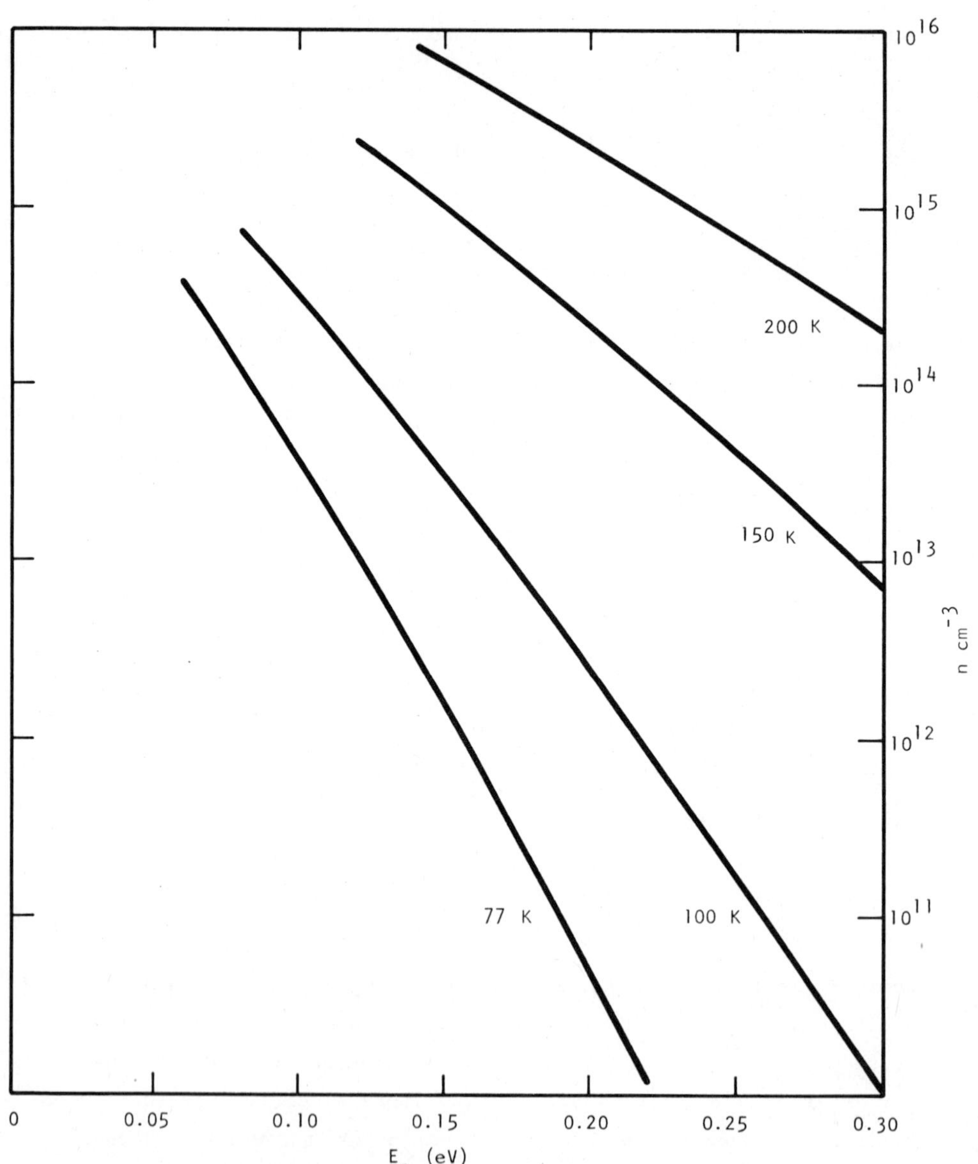

FIG. 10. Calculated intrinsic carrier concentration in (Pb,Sn)Te. (From Ref. 95; reprinted with permission of R. T. Bate.)

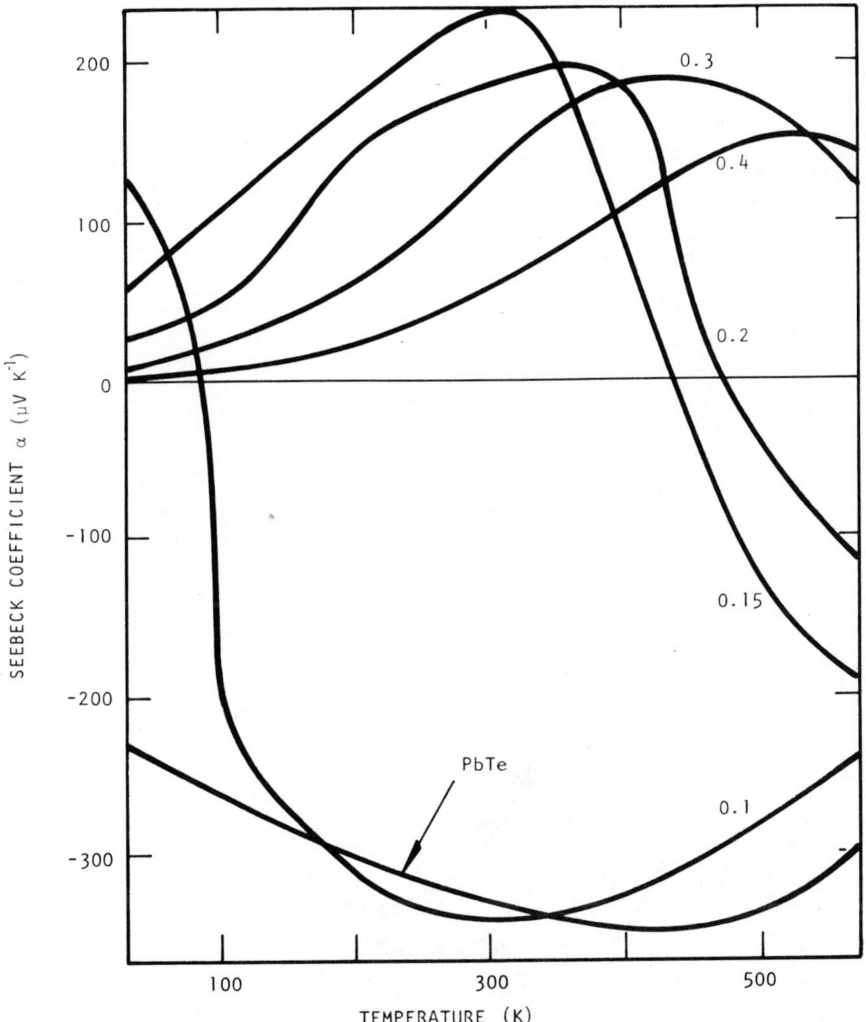

FIG. 11. Temperature dependence of Seebeck coefficient for $Pb_{1-x}Sn_xTe$ alloys. The value of x is designated for each curve. (From Ref. 96; reprinted with permission.)

The crystals were grown from the melt and metal-saturated by annealing $\sim$150°C below the estimated melting point. The data indicate that PbTe is p-type and that $Pb_{0.9}Sn_{0.1}Te$ is n-type up to about 100 K and

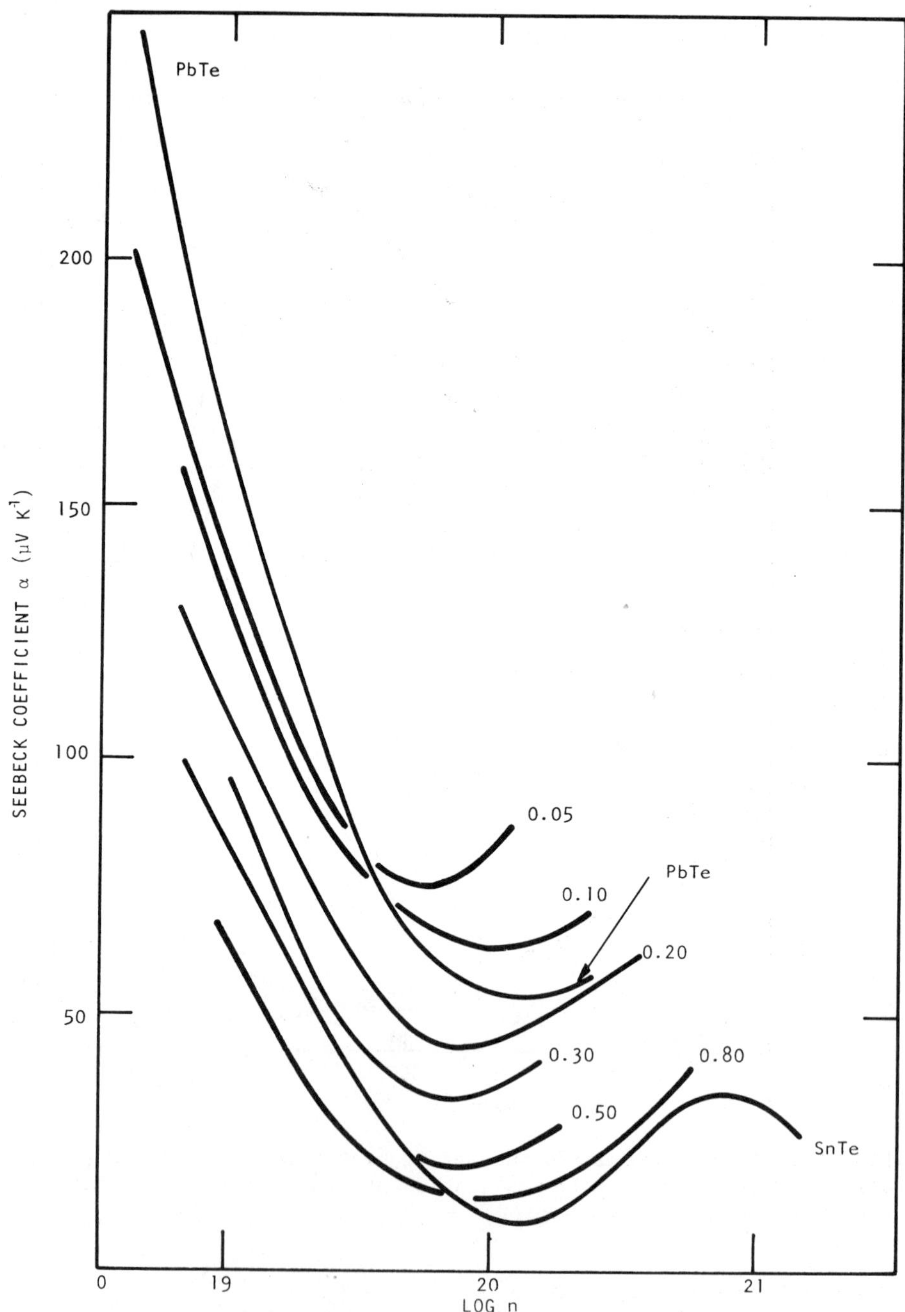

FIG. 12. Dependence of Seebeck coefficient on carrier concentration for $Pb_{1-x}Sn_xTe$ alloys; the curves are labeled with values of x. Measurements at 300 K. (From Ref. 97; reprinted with permission.)

p-type above that temperature. Figure 12 shows the dependence of α
on carrier concentration n from the data of Efimova and Kolomoets
[97]. For all values of x, α decreased to a minimum and then began
to increase as n increased. The authors interpreted this as a contribution from a second valence band, but this interpretation does
not agree with later band models [58,63]. The crystals of (Pb,Sn)Te
of Efimova and Kolomoets were grown by zone melting.

The temperature dependence of Hall coefficient R_H, as determined
by Borde et al. [98], is illustrated in Fig. 13. The temperature-
dependent behavior of the Hall coefficient may be seen to vary with
x, particularly for values of x < 0.4. The Hall coefficient for most
samples first decreased and then increased with increasing temperature;
similar behavior has been noted by others [34,71]. As the value of
x increases, and the hole concentration increases, the Hall coefficient
is less dependent on temperature. Dixon and Bis [63] measured the
temperature dependence of both resistivity and Hall coefficient for
a number of (Pb,Sn)Te films grown by evaporation of PbTe and SnTe onto cleaved NaCl at 300°C; their data are shown in Fig. 14. Since
mobility μ is given by the expression $\mu = R_H/\rho$, the calculated mobility
generally falls with increasing temperature. The change in resistivity
with temperature and composition has been attributed to carrier scattering [63].

Zoutendyk [99] studied Hall mobility in both n- and p-type
$Pb_{0.8}Sn_{0.2}Te$ as a function of carrier concentration; his data are
shown in Fig. 15. His results showed that Hall mobility for both n-
and p-type material at 77 K varies monotonically with carrier concentration over a wide range; he interpreted this as either ionized impurity scattering or mixed ionized impurity/phonon scattering. At
300 K the Hall mobility in both n- and p-type material was independent
of carrier concentrations $<10^{19}$ cm^{-3}, while above this point, the
mobility decreased with increasing carrier concentration. He interpreted these data to mean lattice scattering of carriers at low concentrations and ionized impurity scattering at higher concentrations.
Wagner et al. [100] explained changes in mobility as a function of x

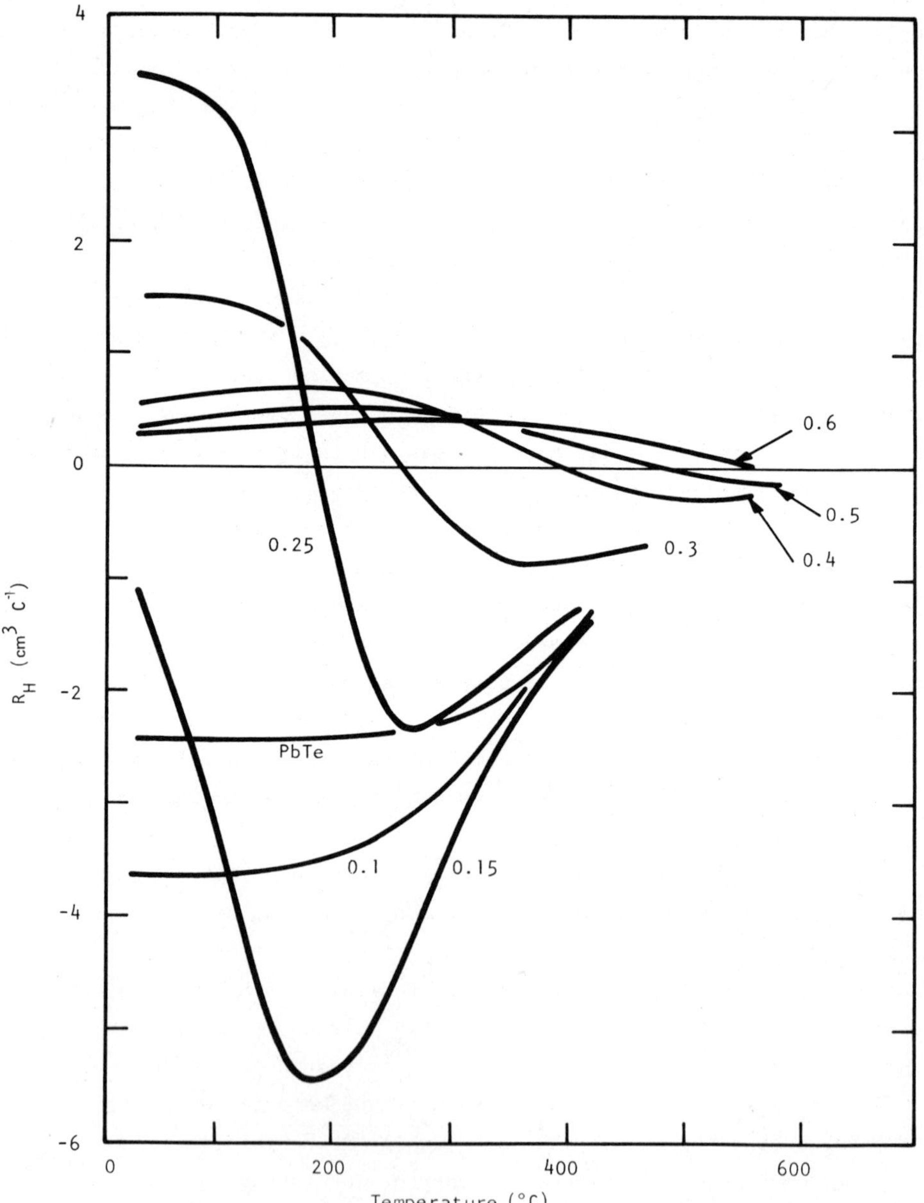

FIG. 13.   Temperature dependence of the Hall coefficient R_H for $Pb_{1-x}Sn_xTe$; the curves are labeled with values of x.   (From Ref. 98; reprinted with permission of the Academy of Sciences, Paris, France.)

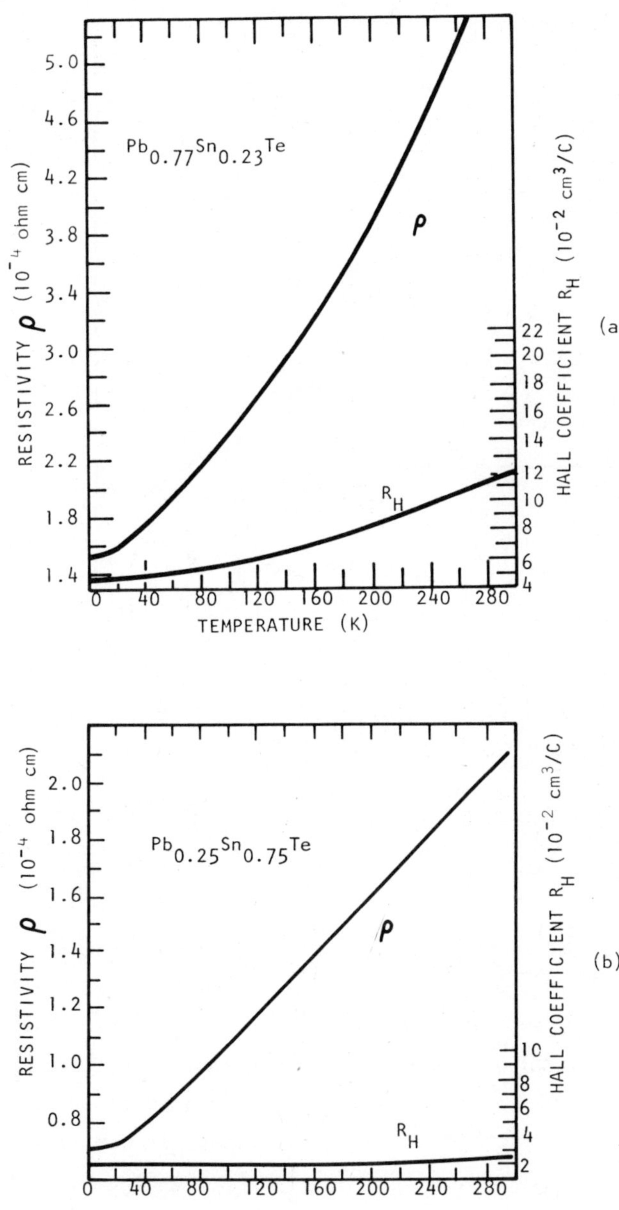

FIG. 14. Temperature dependence of the resistivity ρ and Hall coefficient R_H for bulk Pb$_{1-x}$Sn$_x$Te alloys for different values of x. (From Ref. 63; reprinted with permission.)

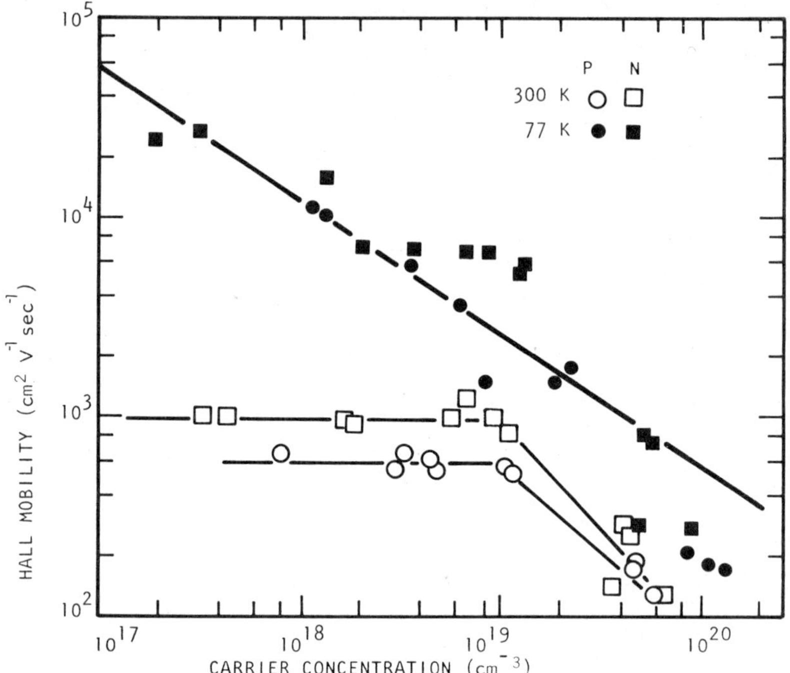

FIG. 15. Carrier concentration and temperature dependence of the Hall mobility in bismuth-doped $Pb_{0.8}Sn_{0.2}Te$. [Reprinted with permission from *Proceedings of the Conference on Physical Semi-Metallic and Narrow Gap Semi-Conductors* (R. T. Bate and D. L. Carter, eds.), P. J. A. Zoutendyk, "Carrier Concentration and Hall Mobility in As-Grown Bismuth-Doped $Pb_{0.8}Sn_{0.2}Te$ Crystals," p. 424. Copyright 1971, Pergamon Press, Ltd.]

in $Pb_{1-x}Sn_xTe$ on the basis of changes in the static dielectric constant and charge carrier effective mass. Ocio [101] concluded from his mobility studies that at low temperatures the carriers are scattered mainly by the short-range potentials of impurities. At room temperature, both impurity scattering and one-phonon acoustic scattering could account for the observed mobility of carrier densities above 10^{19} cm^{-3}. In general, the mobility is proportional to the carrier density as illustrated in Fig. 16 which shows changes in residual mobility versus apparent carrier concentrations for two $Pb_{1-x}Sn_xTe$ alloys. The residual mobility was defined as the mobility

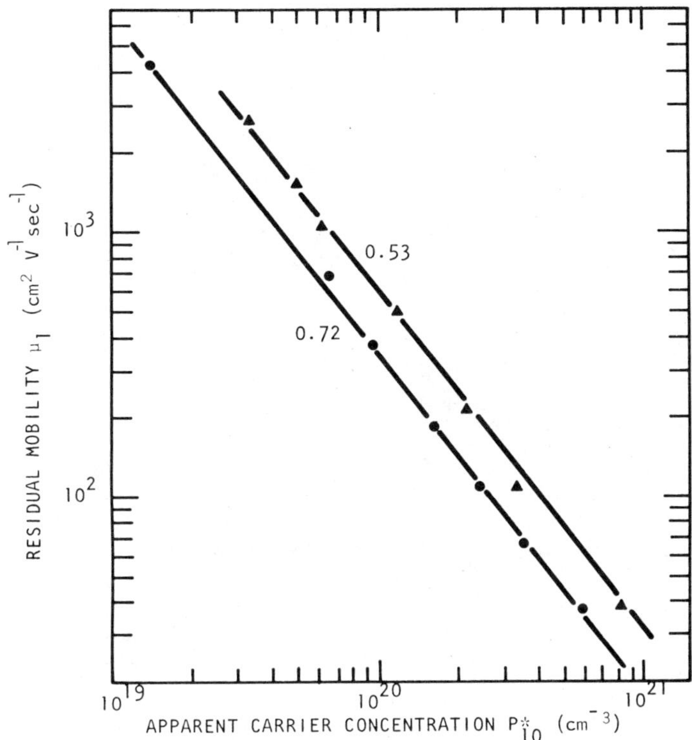

FIG. 16.   Residual mobility vs. apparent carrier concentration (p^*_{10} = $1/eR_{10K}$) for $Pb_{1-x}Sn_xTe$.   Curves are marked to indicate x.   (From Ref. 101; reprinted with permission.)

at 4.2 K and the apparent carrier concentration $p^*_{10} = 1/eR_{10K}$, where R_{10K} is the Hall coefficient at 10 K.   The residual mobility at a given temperature is also dependent on alloy composition [13,57,65,90, 101].   Data measured and collected by Ocio are shown in Fig. 17 for $Pb_{1-x}Sn_xTe$ samples at 77 K with an apparent p-type concentration of 1.4×10^{20} cm^{-3}.

Toneva and Alazhazhian [102] studied carrier mobility in p-type $Pb_{1-x}Sn_xTe$ single crystals for x = 1.00, 0.88, 0.78, and 0.71 at 4.2 to 300 K and carrier concentrations from 1×10^{20} to 1×10^{21} cm^{-3}. The carrier concentration dependence of mobility was calculated based on scattering by impurities and by acoustical and optical phonons.

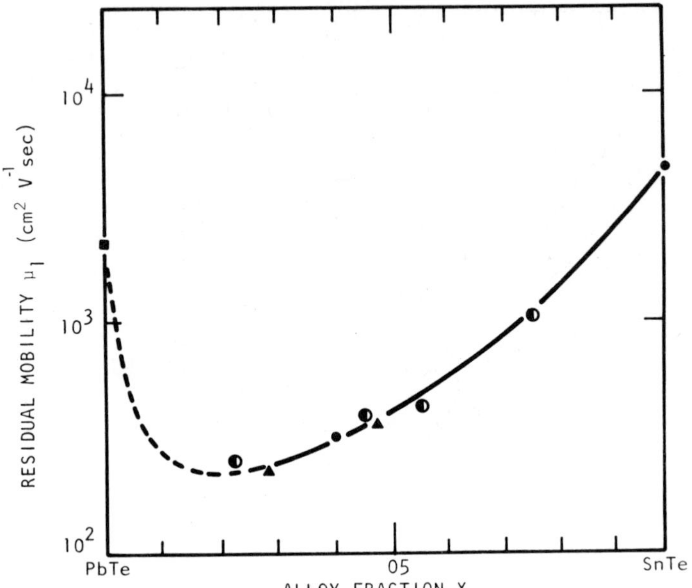

FIG. 17. Residual mobility vs. alloy fraction, for apparent carrier concentration P^*_{10} = 1.4 × 10^{20} cm^3 at 4.2 K. (From Ref. 101; reprinted with permission.)

Impurity scattering is the dominant mechanism at temperatures $\leq$77 K; with increasing temperature the role of acoustical scattering increases.

VII. (Pb,Sn)Te DEVICES

(Pb,Sn)Te has found greatest use as infrared detectors or lasers; a major advantage is that the spectral response peak or emission wavelength can be tailored by varying composition or temperature. In general the photoresponse or emission wavelength is given by

$$\lambda = \frac{hc}{E_g}$$

where h = Planck's constant, c = velocity of light, and E_g = minimum direct bandgap.

Detectors have been operated in the photoconductive model, but the difficulty of preparing low-carrier-concentration ($\leq 10^{16}$ cm^{-3})

uniform material has limited this development. Most detectors have been fabricated to operate in the photovoltaic mode. For these detectors, and lasers as well, device development has depended on the technology for producing high-quality p-n junctions. Junctions have been fabricated by adjusting crystal growth conditions, by annealing in metal-rich or metal-poor ambients, by impurity diffusion, and by producing a Schottky barrier.

In materials which have small effective masses, such as (Pb,Sn)Te, the Burstein shift [103] sets an upper limit on majority carrier concentration in usable material for compositions near zero bandgap. If the majority carrier concentration becomes large enough, the Fermi level shifts into the valence or conduction band and the detector response or laser emission wavelength decreases because the effective bandgap increases. This condition is particularly bad in $Pb_{1-x}Sn_xTe$ alloys grown from an equilibrium melt since as x increases toward material with $E_g = 0$, the carrier concentration in as-prepared material increases. Practically, the direct fabrication of detectors or lasers from a crystal grown from the melt will not give wavelength response or emission greater than 8 to 10 µm; the material must be annealed to a carrier concentration $<10^{18}$ cm^{-3} to avoid the Burstein shift.

Table 4 lists some of the material requirements and properties for commonly used detectors and lasers made from $Pb_{1-x}Sn_xTe$. The necessity of low carrier concentrations and of low operational temperatures limit somewhat the useful optical spectrum of these devices and subsequently the alloy compositions used.

A. Detectors

(Pb,Sn)Te photovoltaic detectors have been produced with high detectivity, negligible power dissipation, high impedance levels, and very low excess (1/f) noise. Detectors are normally operated at 77 K, though detectors with low x, $\sim$0.05, have operated well at 200 K [104]. The detector response speed is moderate and is limited by the capacitance-resistance product because of the high dielectric constant of the material.

TABLE 4

Material Requirements for Commonly Used Detectors and Lasers from $Pb_{1-x}Sn_xTe$

Type of device	Device structure	Operational wavelength (μm)	Alloy composition (100×)	Operational temperature (K)	Carrier type	Number of carriers (cm^{-3} at 300 K)
Detector	p-n Junction	5-17	1.8-27	77-200	n and p	10^{16}-10^{19}
Detector	Schottky barrier	5-17	1.8-27	77-200	n or p	10^{16}-10^{19}
Detector	Photo-conductor	5-17	1.8-27	4.2-77	n or p	10^{16}-10^{17}
Laser	p-n Junction	5-17	1.8-27	4.2-77	n and p	10^{17}-10^{18}
Laser	Optically pumped	5-17	1.8-27	4.2-77	n or p	10^{17}-10^{18}

Corsi [40] produced p-n junctions in (Pb,Sn)Te by a composite sequential sputtering and cosputtering technique on substrates of BaF_2, NaCl, Ge, and Si. Even though some films were single-crystal, peak detectivity D^* was only 2.4×10^8 cm $Hz^{1/2}$ W^{-1} at 77 K for $Pb_{0.83}Sn_{0.17}Te$. Krikorian and Crisp [104] also used sputtering from prereacted (Pb,Sn)Te targets into single-crystal (Pb,Sn)Te substrates to form p-n and n-p heterojunction and homojunction diodes. Two separate layers were deposited with the first layer having a carrier concentration of $<10^{17}$ cm^{-3} and a Hall mobility of $>10^4$ cm^2 V^{-1} sec^{-1} at 77 K. A bias voltage was applied to the substrate during the deposition of the second layer to adjust the type of carrier and its concentration thereby providing the p-n junction for a photovoltaic detector.

Evaporation techniques have been used by several workers [105-110] to provide films for photovoltaic or photoconductive devices on substrates such as PbTe, (Pb,Sn)Te, BaF_2, CaF_2, or NaCl. Generally, annealing was used to reach the desired carrier concentrations and type of carrier. Usually, the film composition was difficult to control and varied in composition during the deposition.

Epitaxial deposition, either liquid-phase or vapor-phase, is the most widely used technique for preparing (Pb,Sn)Te films for photovoltaic detectors, as well as lasers. It is possible to deposit LPE films of low carrier concentration of either conductivity type but, as noted earlier, only p-type low-carrier-concentration films of good quality are possible by VPE. Bellavance and Johnson [42] used In or Pb films to form a detector array of Schottky barriers on p-type $Pb_{0.79}Sn_{0.21}Te$. The metals were deposited on as-grown (Pb,Sn)Te without any surface preparation. A typical response curve of a detector is shown in Fig. 18. The shape of the response curve is nearly that of a quantum detector, and indicates that composition grading in the film is not a problem. There was also good compositional uniformity laterally across the epitaxial film. Element no. 1 of the detector array had the following properties at 77 K:

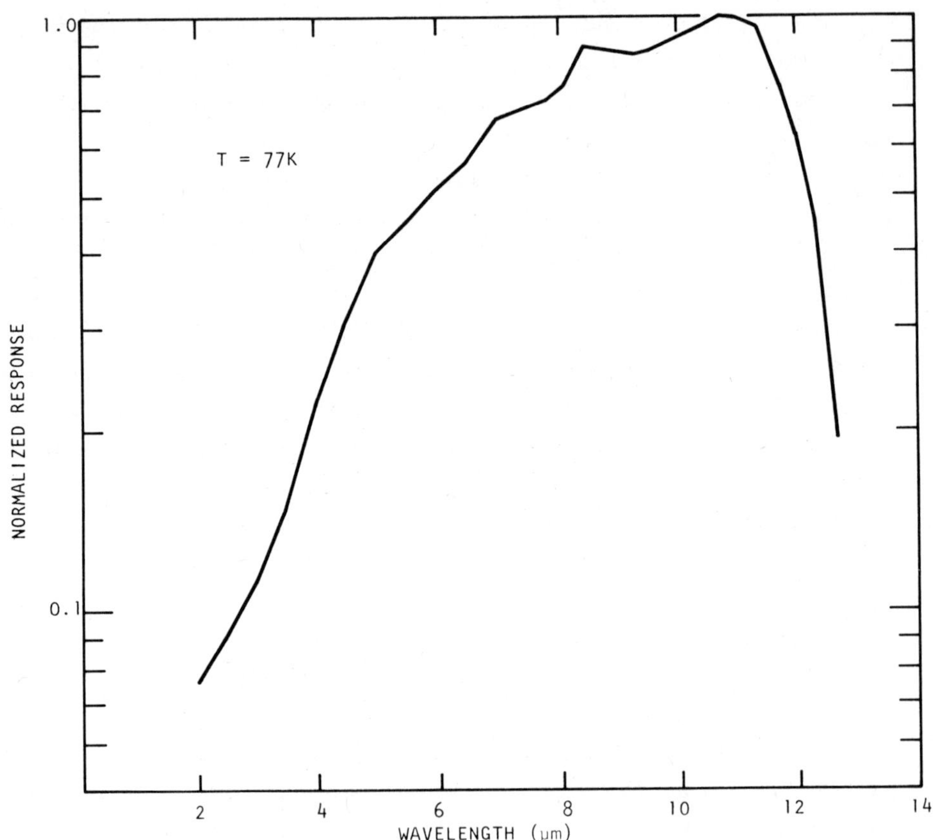

FIG. 18.   Normalized spectral response of a $Pb_{0.79}Sn_{0.21}Te$ VPE detector.   (From Ref. 42; reprinted with permission.)

Field of view	96°
Resistance-area product	0.92 Ω cm^2
Capacitance (zero bias)	0.12 pf
Quantum efficiency	0.53
Detectivity (11, 1050, 1, 96°)	3.9×10^{11} cm Hz$^{1/2}$ W^{-1}

An advantage of this method of material and device fabrication is the very short time required overall.

Several groups [111-115] have used liquid epitaxial methods to grow low-carrier-concentration (Pb,Sn)Te films for infrared detectors.

Either type material with carrier concentrations 10^{15} to 10^{17} cm^{-3} can be grown and used without annealing. Great care must be taken to control the following parameters during LPE: temperature gradient across the growth interface, growth temperature, cooling rate, and crystallographic orientation. Failure to control these variables can lead to irregular surfaces and inclusions. Tomasetta and Fonstad [47] reported good surfaces by LPE using a wiper technique. Wang and Lorenzo [112] made a planar detector array from $Pb_{1-x}Sn_xTe$ deposited on (Pb,Sn)Te by LPE. Impurity diffusion gave them a carrier concentration of 4×10^{16} cm^{-3} in the epitaxial film. At 77 K, the average resistance area product was 1.85 Ω cm^2 for a 124-element array with elements 50 $\times$ 50 μm. At 77 K the average quantum efficiency and peak detectivity at 11 m without an antireflection coating were 0.40 and 2.6×10^{10} cm $Hz^{1/2}$ W^{-1}, respectively, for a 300 K background and a 180° field of view.

Detectors have also been made by diffusion of deposits into bulk p-type (Pb,Sn)Te to form a junction. For long-wavelength detectors, the (Pb,Sn)Te must be annealed to a low carrier concentration to avoid the Burstein shift. Al, In, and Cd readily diffuse into the annealed (Pb,Sn)Te to form p-n junctions, but the position or depth of the p-n junction varies from slice to slice and tends to drift at temperatures over 100°C. When Sb is diffused into p-type (Pb,Sn)Te, it always reaches a concentration of 2 to 5×10^{19} cm^{-3} at the p-n junction [116], independent of hole concentration or Sn content; this concentration is high enough that the detectors showed the Burstein shift. Also, it was noted that the junction diffusion rate increased with decreasing hole concentration in the substrate. Mechanisms which might account for the behavior of Sb diffused into (Pb,Sn)Te include formation of Sb precipitates, compensation of Sb by native defect acceptors, and amphoteric doping behavior. Recently it has been reported that most Cd distributes uniformly as an electrically inactive impurity when Cd diffuses into $Pb_{1-x}Sn_xTe$ and only a small fraction acts as active donors [117-118]. Cd acts as a donor by filling Pb or Sn vacancies.

In general, diffused junction photodiodes have not performed as well as Schottky diodes or as-grown p-n junction detectors. Much remains to be learned about impurity diffusion into (Pb,Sn)Te and about electrical and chemical behavior of these dopants.

B. Lasers

(Pb,Sn)Te lasers have generated considerable interest for use in high-resolution gas spectroscopy and for monitoring gas pollutants. (Pb,Sn)Te laser diodes provide a compact, safe, rugged, and relatively convenient source of tunable infrared radiation of exceptional spectral purity. The spectral purity is advantageous for high-resolution spectroscopy and as a means of eliminating interference from other species while monitoring a specified gas pollutant. Gross tuning is available by varying the composition to change the bandgap [52,119]. Fine tuning may be done by varying the diode temperature at a constant current or by varying current while the diode is held on a constant-temperature heat sink; in either case, the junction temperature changes and hence the bandgap. The diode lasing frequency can also be varied by application of an external magnetic field [120].

Lo [121] made a homojunction (Pb,Sn)Te laser to cover the absorption spectra of such molecules as HCN, NO_2, and SO_2 with absorption frequencies from 1155 to 1380 cm^{-1} as shown in Fig. 19. Figure 20 shows a comparison of temperature tuning and current tuning for one mode.

Kawakami et al. [122], used a $Pb_{0.93}Sn_{0.07}Te$ laser diode with an output power of tens of μW and a linewidth of $<10^{-2}$ cm^{-1} to study the infrared spectrum of ethane, C_2H_6. Several peaks were resolved for the first time. Hinkley [123] used a current-tunable $Pb_{0.88}Sn_{0.12}Te$ diode laser to obtain the absorption spectrum of SF_6 near the P(16) and P(20) CO_2 laser lines at 10.6 μm by both direct and heterodyne techniques. He pointed out that the infrared frequency of such diodes can be tuned continuously over a range much greater than that attainable with a gas laser. The spectrum of NH_3 was obtained over the broad 8.2-10.6-μm region with a single continuous-wave (CW)

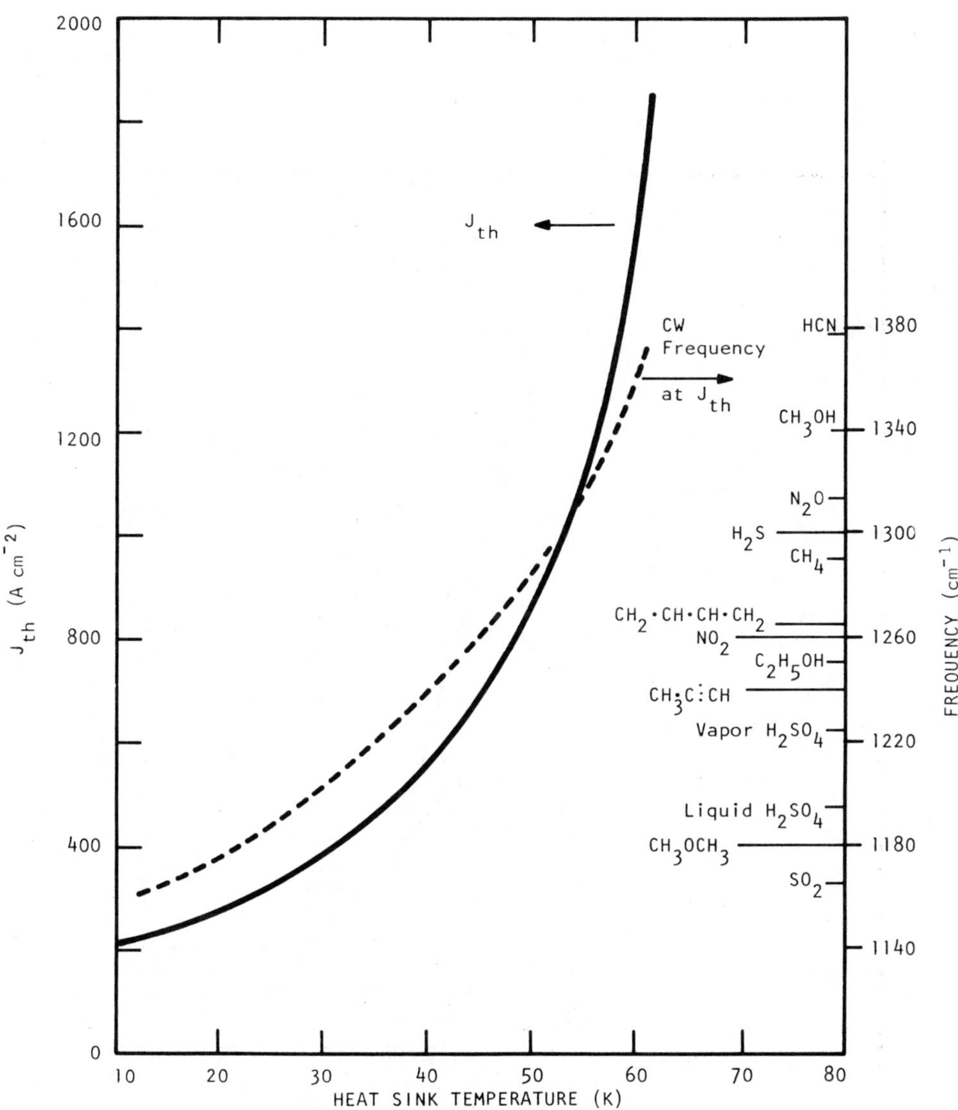

FIG. 19. Dependence of threshold current density (J_{th}) on the heat sink temperature. Also shown are the lasing frequency at J_{th} and some important gases that can be covered with one laser. [From W. O. Lo, "Homojunction Lead Tin-Telluride Diode Lasers," *IEEE J. Quantum Electron.*, *QE-13;* 593 (1977); reprinted with permission.]

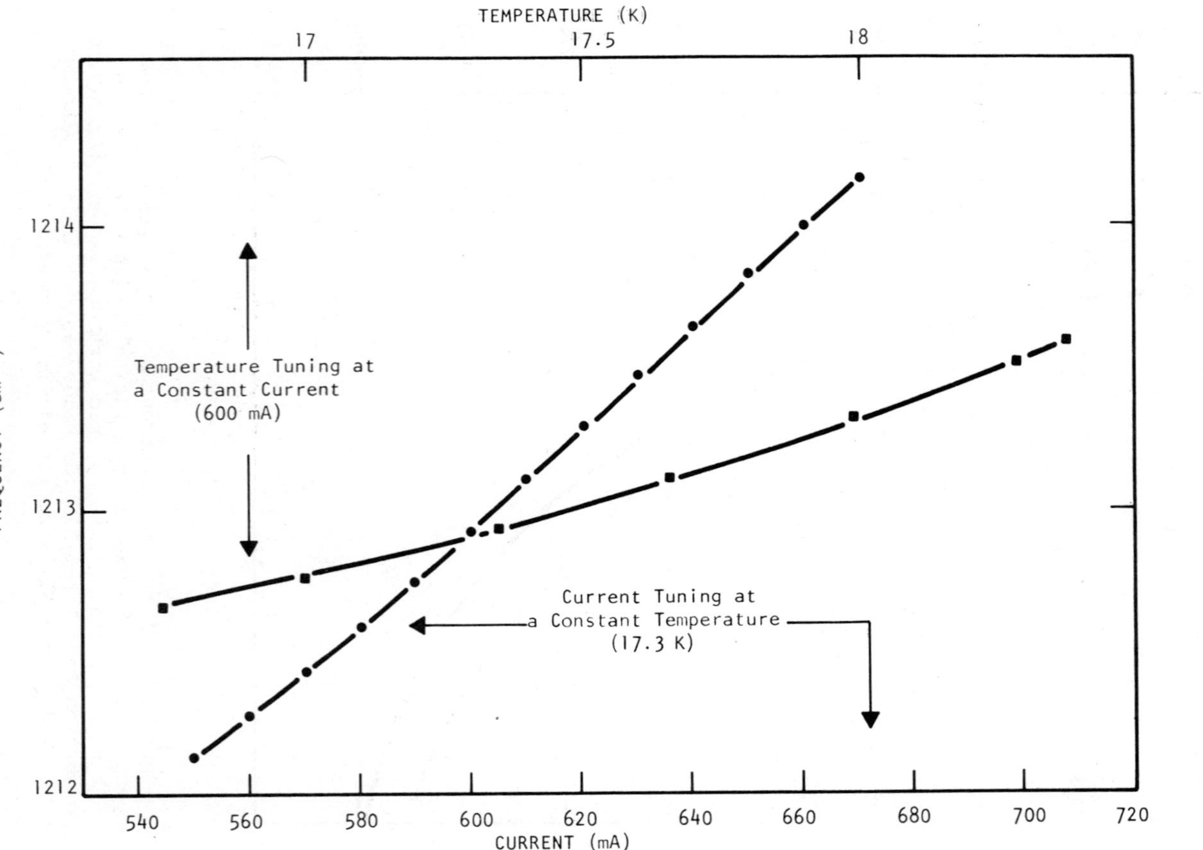

FIG. 20. Comparison of temperature tuning and current tuning of one mode of a (Pb,Sn)Te laser. [From W. Lo, "Homojunction Lead-Tin-Telluride Diode Lasers," *IEEE J. Quantum Electron.*, *QE-13*; 595 (1977); reprinted with permission.]

$Pb_{0.88}Sn_{0.12}Te/PbTe$ laser [124]; the laser had a power >1 mW mode^{-1} and operated CW at >77 K. Antcliffe and coworkers [55,125] used $Pb_{1-x}Sn_xTe$ lasers to obtain high-resolution spectra of NH_3, SF_6, and SO_2. Fine structure was observed in the SO_2 spectrum with separations of 0.05 to 0.1 cm^{-1} between absorption lines. Broadening of the individual lines in SO_2 occurred with increasing SO_2 pressure and with addition of N_2; however, with current-tunable diodes, it was still possible to resolve individual lines under actual monitoring conditions at 1 atm pressure.

The performance of (Pb,Sn)Te lasers is very dependent upon the method of fabrication and on crystalline perfection. The early lasers [25,58] were photodiodes formed by metal-rich annealing, but these lasers showed poor performance, probably because of metal precipitation [120]. Improvements [126] in the optical quality of the end faces by mechanical and etch polishing gave lasers with CW output power over 10 mW and single-longitudinal-mode CW output up to 6 mW at 10.6 µm. Lasers from the same material but with cleaved faces gave peak output powers of a few hundred mW. It was also discovered that laser output power increased as the number of dislocations in the (Pb,Sn)Te decreased [126].

Antcliffe and Wrobel [127] and Lo [121,128] used impurity diffusion to fabricate diode lasers; an advantage of this method is the significantly shorter fabrication time required compared to annealed junctions. Antcliffe and Wrobel used the donor Sb to prepare $Pb_{1-x}Sn_xTe$, lasers with 0.13 < x < 0.20. They vacuum-diffused elemental Sb for several hours at 700°C into p-type (Pb,Sn)Te with a hole concentration of 3 × 10^{19} cm^{-3}. They made both Fabry-Perot (FP) laser diodes (250 to 370 µm between cleaved faces and 250 µm in width) and etched-mesa (EM) structures where the 250-µm square junction was completely defined by an etched surface. Figure 21 shows power output versus diode current at 4.2 K using 1 kHz excitation at 0.1% duty cycle. The maximum CW power was <100 µW at 4.2 K for a single mode.

Lo [128] used Cd diffusion to obtain (Pb,Sn)Te lasers with CW output of 1.25 mW (single-mode) and 2.4 mW (total) at 10.6 µm. These

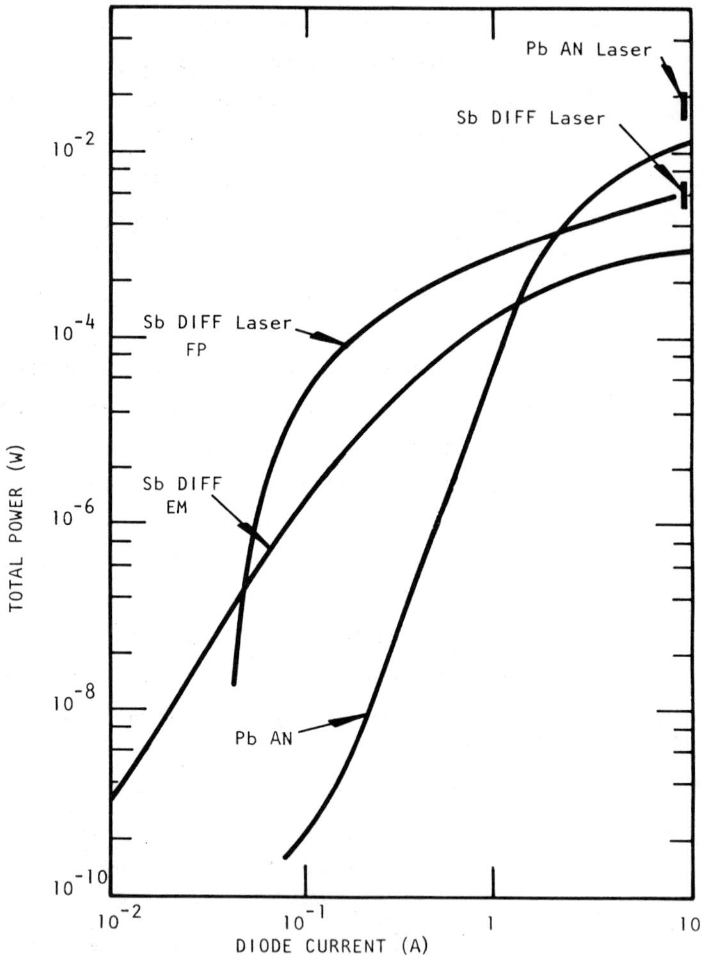

FIG. 21. Total emission versus diode current for typical Sb-doped
and Pb-annealed etched mesa diodes and for FP lasers prepared by Sb
diffusion. The relative performance of lasers prepared by these two
techniques is indicated by vertical bars. (From Ref. 127; reprinted
with permission.)

power levels were attributed to the low-temperature Cd diffusion
(450°C for 1 hr), the low dislocation count ($\leq 10^3$ cm^{-2}), and contact
resistance as low as 3×10^{-5} Ω cm^{-2}. The crystals were grown by
vapor transport with self-nucleation [28]. The low contact resistance

was attributed to better adhesion of the contact caused by a slight
roughening of the surfaces before applying the contacts. However,
lasers fabricated from the same slice often gave different spectral
characteristics; lasers exhibited filamentary multimode spectra,
usually indicative of spatial inhomogeneity. Lo suggested that reducing the cavity width (e.g., by using a stripe geometry) would lead
to single-mode operation. Lo [121] later showed that slow cooling
after crystal growth gave a graded hole carrier concentration in the
surface of the (Pb,Sn)Te crystals. The graded carrier concentration
helped to confine the photons through a waveguiding effect and allowed
CW operation up to 61 K.

Donnelly et al. [129] used proton bombardment of Bridgman grown
crystals to fabricate diode lasers in $Pb_{0.88}Sn_{0.12}Te$. These lasers
operated CW at 4.2 K with a CW threshold of about 600 A cm^{-2}. The
output emission spectrum was multimode with output energy distributed
among several modes between 10.47 and 10.56 μm. The advantage of
this process, which has not been fully exploited, is that the photodiodes can be made planar and uniform so that large areas could poss-
ibly be processed as stripe geometry lasers.

Ralston et al. [130] produced stripe geometry (Pb,Sn)Te diode
lasers from vapor-grown crystals by using photoresist for patterning
and ion milling to shape the patterns. These lasers emitted in a
fundamental spatial transverse mode which reduced the number of para-
sitic modes produced. The emission consisted of regularly spaced
modes corresponding to a longitudinal cavity with an effective mode
number of 7.1; the measured width of the emitting region corresponded
to the junction width in both 50-μm and 100-μm wide stripe lasers.
The junction depth was ∿50 μm for all lasers. Antireflection coat-
ings produced as much as a fourfold increase in output power.

Heterostructures have been used to produce lasers [47,131,132];
the geometry provides waveguiding due to optical confinement between
layers. (Pb,Sn)Te heterostructures have been grown by liquid phase
[47,133,134] and by molecular beam epitaxy [131,132,135] to form
layered structures such as n-(Pb,Sn)Te (epi)/p-PbTe (substrate) and

n-PbTe (epi)/n-(Pb,Sn)Te (epi)/PbTe (substrate). Lasing occurred in
the (Pb,Sn)Te layers and indicated that the presence of heterojunc-
tions does not introduce a significant number of new nonradiative
recombination centers and that efficient minority carrier injection
occurs across the heterojunction. It was shown [130] experimentally
and calculated [136] that with a double heterojunction geometry a
fourfold reduction in pulsed threshold can be obtained over that of
a comparable single heterojunction laser; these films had a 4-μm
thick middle region.

Groves et al. [133] inhibited junction diffusion during LPE
growth of double heterostructure $Pb_{0.88}Sn_{0.12}Te$ lasers by use of Tl-
doped PbTe substrates. They obtained CW operation up to 80 K. The
CW threshold current density increased from 1.6×10^3 A cm^{-2} at 12 K
to 4.2×10^3 A cm^{-2} at 77 K. The measured CW output power was 10 mW
in four modes at 12 K and 1.2 mW in a single mode at 77 K

Distributed feedback (Pb,Sn)Te double heterostructure lasers
have been operated in stripe geometry, both pulsed [136] and CW [137].
The pulsed lasers gave only one or two modes and thus are particularly
significant for tunable local oscillator applications. The heterostructures were made by molecular beam techniques and the stripe geo-
metries were made by outlining the patterns with photoresist followed
by ion milling. The grating of 1.1-μm periodicity operated in the
first Bragg order. Striped mesas 32 μm wide were formed with electrically active regions 450 μm in length at the output end of the
mesas and electrically insulated regions of comparable length at the
other end to spoil the Fabry-Perot cavity. It was possible to achieve
a large coupling coefficient for Bragg reflection without locating
the grating at a hetero interface or using multilayered separate confinement structures. These workers found continuous current tuning
of mode frequency by varying both the diode current and heat sink
temperature; this is illustrated in Figs. 22(a) and (b).

Figure 22(a) shows how the mode frequency varies with temperature
for distributive feedback (DFB) lasers as compared to FP-type lasers.
Only two modes were obtained for the DFB lasers and their frequency

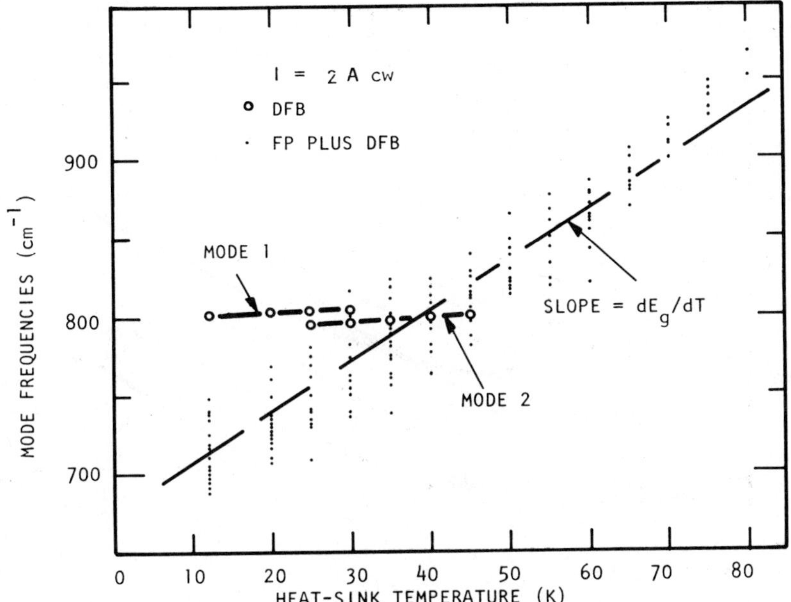

FIG. 22(a). Mode frequencies of a PbSnTe DFB diode laser at 2 A
compared to the mode frequencies of a device having both DFB and
reflective feedback (labeled FP plus DFB) as a function of heat sink
temperature. The dashed line with slope labeled $dE_g dT$ shows the temperature dependence of the spontaneous emission. At 2 A only modes
1 and 2 are seen at any temperature for the DFB laser. (From Ref.
137; reprinted with permission.)

varied only slightly with temperature as opposed to FP-type lasers.
Figure 22(b) shows how the frequency of DFB lasers varies with diode
current and that by using a combination of diode current and heat
sink temperature, one can obtain modes covering a wide range of
frequency.

VIII. SUMMARY

(Pb,Sn)Te is a semiconductor which can be prepared in a high degree
of crystalline perfection in large single crystals by vapor transport
techniques. Epitaxial films of high perfection can be grown by either
vapor-phase or liquid-phase methods. However, defects such as voids,

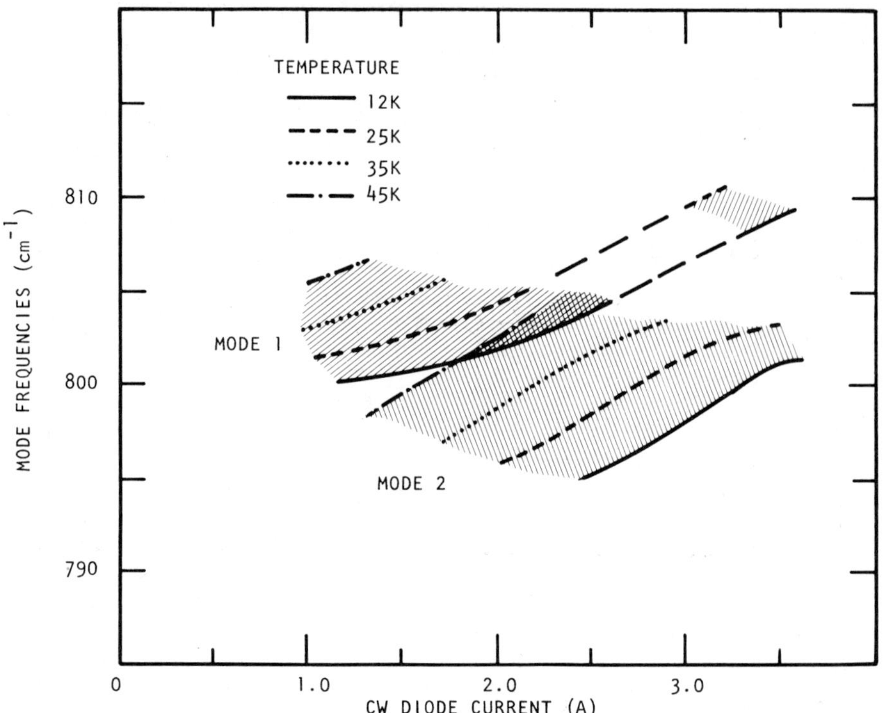

FIG. 22(b). Frequencies of the two lowest thredhold modes of a
PbSnTe DFB laser versus diode current with heat sink temperature varying as indicated between 12 and 45 K. The two-parameter (current-
temperature) tuning range of mode 1 consists of two regions with a
gap in the emission indicated by the dashed lines. (From Ref. 137;
reprinted with permission.)

inclusions, and dislocations occur easily unless great care is taken

during crystal growth and handling. Because of the thermodynamic

properties of the PbTe-SnTe system, (Pb,Sn)Te as prepared is usually

nonstoichiometric with a large concentration of p-type defects. The

carrier defect concentration can be varied by selection of growth

temperature, annealing procedures, change of composition, or incorpora-

tion of electrically active dopants.

The bandgap of $Pb_{1-x}Sn_xTe$ is direct but varies with both composi-

tion and temperature. As the value of x increases the bandgap de-

creases until zero bandgap is reached. With further increase in x, bandgap again increases. This change in bandgap with composition has been explained by a band inversion model based on relativistic effects in Pb and Sn. Data from experiments tend to confirm the basic concepts of this bandgap model.

The availability of high-quality (Pb,Sn)Te with its small direct bandgap has led to application studies of detectors and lasers. Frequency (wavelength) tuning is possible by varying composition; in addition, laser frequency may be tuned by current density and operating temperature. Power output of the lasers is related to the number of dislocations present, to the geometry of the laser cavity, to the carrier concentration, and to the operating temperature. $Pb_{1-x}Sn_xTe$ lasers have been used for spectroscopic studies of various compounds and for the detection of numerous pollutants.

The preparation and properties of $Pb_{1-x}Sn_xTe$ has been a fruitful area for study. These studies have indicated a potential usage in infrared detector and lasers.

REFERENCES

1. N. K. Abrikosov, K. A. Duldina, and F. A. Dasilyan, *Zh. Neorgan. Khim.*, *3*, 1632 (1958).

2. R. Mazelsky, M. S. Lubell, and W. E. Kramer, *J. Chem. Phys.*, *37*, 45 (1962).

3. A. M. Reti, A. K. Jena, and M. B. Bever, *Trans. Met. Soc. AIME*, *242*, 371 (1968).

4. A. R. Calawa, T. C. Harman, M. Finn, and P. Youtz, *Trans. Met. Soc. AIME*, *242*, 374 (1968).

5. I. Melngailis and T. C. Harman in *Semiconductors and Semimetals*, Vol. 5 (R. K. Willardson and A. C. Beer, eds.), Academic, New York, 1970, p. 115.

6. A. Laugier, J. Cadoz, M. Faure, and M. Moulin, *J. Cryst. Growth*, *21*, 235 (1974).

7. E. Miller and K. L. Komarek, *Trans. Met. Soc. AIME*, *236*, 832 (1966).

8. K. J. Linden and C. A. Kennedy, *J. Appl. Phys.*, *40*, 2595 (1969).

9. J. S. Harris, J. F. Longo, E. R. Gertner, and J. E. Clark, *J. Cryst. Growth*, *28*, 334 (1975).

10. J. W. Wagner and R. K. Willardson, *Trans. Met. Soc. AIME*, *242*, 366 (1968).

11. R. F. Brebrick, *J. Phys. Chem. Solids*, *32*, 551 (1971).

12. R. F. Bis and J. R. Dixon, *J. Appl. Phys.*, *40*, 1918 (1969).

13. J. W. Wagner and J. C. Woolley, *Mater. Res. Bull.*, *2*, 1055 (1967).

14. N. R. Short, *Brit. J. Appl. Phys.*, *D1*, 129 (1968).

15. G. Dionne and J. C. Woolley, *J. Electrochem. Soc.*, *119*, 784 (1972).

16. J. F. Butler and T. C. Harman, *J. Electrochem. Soc.*, *116*, 260 (1969).

17. M. K. Norr, *J. Electrochem. Soc.*, *109*, 433 (1962).

18. J. W. Wagner and R. K. Willardson, *Trans. Met. Soc. AIME*, *245*, 461 (1969).

19. H. Kimura, *J. Electron. Mater.*, *1*, 166 (1972).

20. D. A. Northrop, *J. Electrochem. Soc.*, *118*, 1365 (1971).

21. D. A. Northrop, *J. Phy. Chem.*, *75*, 118 (1971).

22. E. E. Hansen and Z. A. Munir, *J. Electrochem. Soc.*, *118*, 983 (1971).

23. R. Colin and J. Drowart, *Trans. Faraday Soc.*, *60*, 673 (1964).

24. S. E. R. Hiscocks, *J. Crystal Growth*, *17*, 222 (1972).

25. J. F. Butler and T. C. Harman, *Appl. Phys. Lett.*, *12*, 347 (1968).

26. O. Ohtsuki, K. Shinohara, and O. Ryuzan, *Jap. J. Appl. Phys.*, *10*, 515 (1971).

27. R. K. Pandey, *Solid State Commun.*, *15*, 449 (1974).

28. W. Lo, *J. Electron. Mater.*, *6*, 39 (1977).

29. W. Lo, G. P. Montgomery, and D. E. Swets, *J. Appl. Phys.*, *47*, 267 (1976).

30. T. C. Harman and J. P. McVittie, *J. Electron. Mater.*, *3*, 843 (1974).

31. S. G. Parker, J. E. Pinnell, and R. E. Johnson, *J. Electron. Mater.*, *3*, 731 (1974).

32. S. G. Parker, *J. Electron. Mater.*, *5*, 497 (1976).

33. N. Tamari and H. Shtrikman, *J. Cryst. Growth*, *43*, 378 (1977).

34. E. G. Bylander, *Mater. Sci. Eng.*, *1*, 190 (1966).

35. E. Holloway, E. M. Logothetis, and E. Wilkes, *J. Appl. Phys.*, *41*, 3543 (1970).

36. D. P. Mathur, *Opt. Eng.*, *14*, 351 (1975).

37. R. F. Bis, J. R. Dixon, and J. R. Lowney, *J. Vac. Sci. Tech.*, *9*, 226 (1972).

38. J..R. Lowney and D. A. Cammack, *Mater. Res. Bull.*, *9*, 1639 (1974).

39. I. Kasai, J. Hornung, and J. Baars, *J. Electron. Mater.*, *4*, 311 (1975).

40. C. Corsi, *J. Appl. Phys.*, *45*, 3467 (1974).

41. W. Rolls, R. Lee, and R. J. Eddington, *Solid State Electron.*, *13*, 75 (1970).

42. D. W. Bellavance and M. R. Johnson, *J. Electron. Mater.*, *5*, 363 (1976).

43. S. G. Parker, *J. Electrochem. Soc.*, *123*, 920 (1976).

44. D. M. Freik, B. V. Vasilishin, K. V. Tsidylo, V. V. Prokopiv, and A. I. Grushin, *Fiz. Elektron*, *14*, 57 (1977).

45. J. F. Longo, J. S. Harris, E. R. Gertner, and J. C. Chu, *J. Cryst. Growth*, *15*, 107 (1972).

46. J. F. Longo, E. R. Gertner, and J. S. Harris, *J. Nonmet.*, *1*, 321 (1973).

47. L. R. Tomasetta and C. G. Fonstad, *Appl. Phys. Lett.*, *24*, 567 (1974).

48. S. H. Groves, *J. Electron. Mater.*, *6*, 195 (1976).

49. M. G. Astles, C. Pickering, and M. L. Young, *J. Electron. Mater.*, *8*, 603 (1979).

50. L. Ickert, L. Doeweritz, M. Hadan, and M. Muehlberg, *Z. Wiss, Math-Naturiss. Reihe*, *25*, 341 (1976).

51. T. C. Harman, *J. Nonmet.*, *1*, 183 (1973).

52. C. R. Hewes, M. S. Adler, and S. D. Senturia, *J. Appl. Phys.*, *44*, 1327 (1973).

53. J. F. Longo, E. R. Gertner, and A. S. Joseph, *Appl. Phys. Lett.*, *19*, 202 (1971).

54. L. R. Tomasetta and C. G. Fonstad, *Mater. Res. Bull.*, *9*, 799 (1974).

55. G. A. Antcliffe and J. S. Wrobel, *Appl. Optics*, *11*, 1548 (1972).

56. K. J. Linden, *J. Electrochem. Soc.*, *120*, 1131 (1973).

57. S. G. Parker and R. L. Guldi, *J. Electrochem. Soc.*, *121*, 90C (1974).

58. J. O. Dimmock, I. Melngailis, and A. J. Strauss, *Phys. Rev. Lett.*, *16*, 1193 (1966).

59. J. F. Butler and A. R. Calawa in *Physics of Quantum Electronics* (P. L. Kelley, B. Lox, and P. E. Fannerwald, eds.), McGraw-Hill, New York, p. 458.

60. P. M. Nikolic, *Brit. J. Appl. Phys.*, *1*, 190 (1966).

61. L. Esaki and P. J. Stiles, *Phys. Rev. Lett.*, *16*, 1075 (1965).

62. K. Takasaki and S. Tanaka, *Phys. Stat. Sol.*, *40*, 173 (1977).

63. J. R. Dixon and R. F. Bis, *Phys. Rev.*, *176*, 942 (1968).

64. E. A. Gurieva, I. N. Dubrovskaya, and B. A. Efimova, *Fiz. Tekh. Poluprov.*, *4*, 245 (1970).

65. D. M. Freik, B. F. Bilen'kii, I. I. Brodyn, V. V. Voitkiv, M. I. Belei, and M. A. Gaushchak, *Tezisy. Dokl.-5th Vses. Konf. Khim. Suyazi Popuprovodn. Polumetallakh*, 112 (1974).

66. V. M. Walz, U. S. Nat. Tech. Inform. Serv., AD Rep. No. 751650, 1972.

67. J. O. Dimmock, *Proc. Conf. Phys. Semi-Met. and Narrow-Gap Semicond.*, Dallas, Texas, March 20-21, 1970 (D. L. Carter and R. F. Bate, eds.), Pergamon, New York, 1971, p. 319.

68. J. B. Conklin, Jr., L. E. Johnson, and G. W. Pratt, Jr., *Phys. Rev.*, *137*, A1282 (1965).

69. F. Herman and S. Skillman, *Atomic Structure Calculations*, Prentice-Hall, Englewood Cliffs, N.J., 1963.

70. M. S. Adler, C. R. Hewes, and S. D. Senturia, *Phys. Rev. B*, *7*, 5186 (1973).

71. C. R. Hewes, M. S. Adler, and S. D. Senturia, *Phys. Rev. B*, *7*, 5195 (1973).

72. R. S. Allgaier and P. O. Scheie, *Bull. Amer. Phys. Soc.*, *6*, 436 (1961).

73. R. S. Allgaier and B. B. Houston, Jr., *Proc. Int. Conf. Phys. Semicond.*, Institute of Physics and Physical Society, London, 1962, p. 172.

74. B. B. Houston and R. S. Allgaier, *Bull. Amer. Phys. Soc.*, *9*, 293 (1964).

75. J. A. Kafalas, R. F. Brebrick, and A. J. Strauss, *Appl. Phys. Lett.*, *4*, 93 (1964).

76. B. A. Efimova and L. A. Kolomoets, *Fiz. Tverd. Tela*, *7*, 424 (1965); transl.: *Soviet Phys.-Solid State*, *7*, 339 (1965).

77. J. R. Burke, Jr., R. S. Allgaier, B. B. Houston, Jr., J. Babiskin, and P. G. Siebenmann, *Phys. Rev. Lett.*, *14*, 360 (1965).

78. B. A. Efimova, V. I. Kaidanov, B. Ya. Moizhes, and I. A. Chernik, *Fiz. Tverd. Tela*, *7*, 2524 (1965); transl.: *Soviet Phys.-Solid State*, *7*, 2032 (1966).

79. H. Kohler, *Z. Angew. Phys.*, *4*, 270 (1967).

80. R. N. Tauber and I. B. Cadoff, *J. Appl. Phys.*, *38*, 3714 (1967).

81. D. H. Damon, C. R. Martin, and R. C. Miller, *J. Appl. Phys.*, *34*, 3083 (1963).

82. F. Seddon, S. C. Gupta, C. Isci, and G. A. Saunders, *J. Mater. Sci.*, *11*, 1756 (1976).

83. D. M. Korn and R. Braunstein, *Phys. Stat. Sol.*, *B*, *50*, 77 (1972).

84. P. Drude, *Ann. Phys.* (Leipzig), *14*, 936 (1904).

85. J. R. Dixon and H. R. Riedl, *Phys. Rev.*, *138*, A387 (1965).

86. R. F. Bis and J. R. Dixon, *Phys. Rev.*, B2, 1004 (1970).

87. J. R. Burke and H. R. Riedl, *Phys. Rev.*, *184*, 830 (1969).

88. S. Rabil, *Phys. Rev.*, *182*, 821 (1969).

89. Y. W. Tsang and M. L. Cohen, *Phys. Rev.*, B3, 1254 (1971).

90. B. Ellis and F. S. Moss, *Proc. 3rd Int. Photocond. Conf.*, London, 1969 (E. M. Pell, ed.), Pergamon, Oxford, 1971, p. 211.

91. F. F. Sizov, G. V. Lashkarev, V. B. Orletskii, and E. T. Grigorovich, *Fiz. Tekh. Pologprovodn.*, *8*, 2074 (1974).

92. J. R. Lowrey and S. D. Senturia, *J. Appl. Phys.*, *47*, 1771 (1976).

93. S. Takano, S. Hotta, H. Kawamara, Y. Kato, K. Kobayoshi, and K. Komatoubara, *Bussei*, *15*, 81 (1974).

94. R. T. Bate, D. L. Carter, and J. S. Wrobel, *Phys. Rev. Lett.*, *25*, 159 (1970).

95. R. T. Bate, Texas Instruments, Dallas, Texas, unpublished data.

96. A. A. Machonis and I. B. Cadoff, *Trans. TMS-AIME*, *230*, 333 (1964).

97. B. A. Efimova and L. A. Kolomoets, *Sov. Phys.-Solid State*, *7*, 339 (1965).

98. D. Borde, H. J. Albany, M. Roudier, and M. Vandevyver, *Comp. Rend.*, *Ser. B*, *260*, 5235 (1966).

99. P. J. A. Zoutendyk, *Proc. Conf. Phys. Semi-Met. and Narrow-Gap Semicond.*, Dallas, Texas, March 20-21, 1970 (R. T. Bate and D. L. Carter, eds.), Pergamon, New York, 1971, p. 421.

100. J. W. Wagner, A. G. Thompson, and R. K. Willardson, *J. Appl. Phys.*, *42*, 2515 (1971).

101. M. Ocio, *Phys. Rev.*, *B*, *10*, 4274 (1974).

102. A. Toneva and A. Alazhazhian, *Phys. Stat. Sol. A*, *44*, 621 (1977).

103. E. Burstein, *Phys. Rev.*, *93*, 632 (1954).

104. E. Krikorian and M. J. Crisp, U. S. Patent 4,057,476.

105. E. M. Logothetis and H. Holloway, *J. Appl. Phys.*, *43*, 256 (1972).

106. A. Nucciotti, P. DeStefano, P. Masheretti, and G. Samoggia, *Phys. Stat. Sol.*, *12*, 193 (1972).

107. C. Corsi, I. Alfieri, and G. Petrocco, *Appl. Phys. Lett.*, *24*, 484 (1974).

108. K. P. Scharnhorst, R. F. Bis, J. R. Dixon, B. B. Houston, Jr., R. W. Brown, and H. R. Riedl, U. S. Patent 3,961,998.

109. C. Corsi, A. D'Amico, G. Petrocco, G. Fainelli, G. Coppuccio, and G. Vitali, *Thin Solid Films*, *36*, 239 (1976).

110. W. H. Rolls and C. V. Eddolls, *Infrared Phys.*, *13*, 143 (1973).

111. A. H. Lockwood, J. R. Balon, P. S. Chia, and F. J. Renda, *Infrared Phys.*, *16*, 509 (1976).

112. C. C. Wang and J. S. Lorenzo, *Infrared Phys.*, *17*, 83 (1977).

113. M. Moulin, M. Faure, and J. L. Cadoz, *Ann. Chem.*, *9*, 103 (1974).

114. C. C. Wang and S. R. Hampton, *Solid State Electron.*, *18*, 121 (1975).

115. A. M. Andrews, J. A. Higgins, J. F. Longo, E. R. Gertner, and J. G. Pasko, *Appl. Phys. Lett.*, *21*, 285 (1972).

116. R. L. Guldi and G. A. Antcliffe, *J. Electrochem. Soc.*, *121*, 1523 (1974).

117. E. Silberg and A. Zemel, *J. Electron. Mater.*, *8*, 99 (1979).

118. H. R. Vydyanath, *J. Appl. Phys.*, *47*, 5003 (1976).

119. J. O. Dimmock, I. Melngailis, and A. J. Strauss, U. S. Patent 3,748,593.

120. G. A. Antcliffe and S. G. Parker, *J. Appl. Phys.*, *44*, 4145 (1973).

121. W. Lo, *IEEE J. Quantum Electron.*, *QE-13*, 591 (1977).

122. H. Kawakami, Y. Izawa, and C. Yamanaka, *Jap. J. Appl. Phys.*, *17*, 461 (1978).

123. E. D. Hinkley, *Appl. Phys. Lett.*, *16*, 351 (1970).

124. K. W. Nill, S. H. Groves, and A. J. Strauss, *IEEE J. Quantum Electron.*, *10*, 792 (1974).

125. G. A. Antcliffe, R. F. Bate, S. G. Parker, and J. S. Wrobel, 142 Nat. Mtg. Electrochem. Soc., Houston, Texas, Oct. 1972, Extd. Abst. 205, Electrochemical Society, Princeton, N.J., p. 521.

126. J. N. Walpole, A. R. Calawa, R. W. Ralston, and T. C. Harman, *J. Appl. Phys.*, *44*, 2905 (1973).

127. G. A. Antcliffe and J. S. Wrobel, *Appl. Phys. Lett.*, *17*, 290 (1970).

128. W. Lo, *Appl. Phys. Lett.*, *28*, 154 (1976).

129. J. P. Donnelly, A. R. Calawa, T. C. Harman, A. G. Foyt, and W. T. Lindley, *Solid State Electron.*, *15*, 403 (1972).

130. R. W. Ralston, I. Melngailis, A. R. Calawa, and W. F. Lindley, *IEEE J. Quantum Electron.*, *9*, 350 (1973).

131. L. R. Tomasetta and C. G. Fonstad, *Appl. Phys. Lett.*, *25*, 440 (1974).

132. R. W. Ralston, J. N. Walpole, T. C. Harman, and I. Melngailis, *Appl. Phys. Lett.*, *26*, 64 (1975).

133. S. H. Groves, K. W. Nill, and A. J. Strauss, *Appl. Phys. Lett.*, *25*, 331 (1974).

134. L. N. Kurbatov, A. D. Britov, S. M. Karavaev, Yu. I. Gorina, G. A. Kalyuzhnaya, and P. M. Starik, *Kvantovaya Elektron.*, *2*, 2084 (1975).

135. L. N. Kurbatov, A. D. Britov, S. M. Kalyuzhnaya, and T. F. Terekhovich, *Kvantovaya Elektron.*, *4*, 428 (1975).

136. J. N. Walpole, A. R. Calawa, S. R. Chinn, S. H. Groves, and T. C. Harman, *Appl. Phys. Lett.*, *29*, 307 (1976).

137. J. N. Walpole, A. R. Calawa, S. R. Chinn, S. H. Groves, and T. C. Harman, *Appl. Phys. Lett.*, *30*, 524 (1977).

Chapter 2

BASIC CRYSTALLIZATION PROCESSES IN SILVER HALIDE PRECIPITATION

Jong S. Wey

Research Laboratories
Eastman Kodak Company
Rochester, New York

I. INTRODUCTION

The basic component of conventional, silver-based photographic materials is the silver halide emulsion, a dispersion of minute silver halide crystals (about 0.03 to 3 μm in size) in a protective colloid. To prepare the light-sensitive dispersion, the silver halide crystals are precipitated by the reaction of an aqueous alkali halide (chloride, bromide, or iodide) solution and an aqueous silver solution (commonly silver nitrate). In any practical application of this reaction, the presence of a peptizing and protective colloid is required in order to give a stable dispersion of the silver halide crystals without the occurrence of clumping. In most instances gelatin is used as the protective colloid at a concentration of 0.5 to 8 wt% [1-3]. Several crystallization processes may occur during this complex precipitation [4-6]. This chapter provides a concise picture of the current knowledge of these processes. Earlier reviews in this area are available in Refs. 1 and 4.

A. Solubility of Silver Halides

Silver halide crystals are sparingly soluble substances. The activities of the silver and halide ions in solution are related by the activity solubility product K_{sp}:

$$[Ag^+][X^-] = K_{sp} \tag{1}$$

where $X^- = Cl^-$, Br^-, or I^-. In the case of AgX precipitations with low ionic strength, the activity coefficients of the ions in solution are about unity and an approximate value of the solubility product can be obtained by using molar concentrations instead of activities. The solubility product principle is frequently employed in this approximate form. At 25°C, the K_{sp} values for AgCl, AgBr, and AgI are $10^{-9.8}$, $10^{-12.3}$, and $10^{-16.1}$, respectively [1].

The silver ion concentration is an important quantity in AgX emulsion chemistry and is normally expressed by

$$pAg = -\log[Ag^+] \tag{2}$$

The pAg can be measured with a silver electrode and an Ag/AgCl reference electrode pair. By analogy with the definition of pAg, pX and
pK_{sp} are defined as $-\log[X^-]$ and $-\log[K_{sp}]$, respectively. Thus, Eq.
(1) becomes

$$pAg + pX = pK_{sp} \tag{3}$$

When $pAg = pX = pK_{sp}/2$, the system is at the so-called equivalence
point. At lower pAg values there is an excess of silver ions, and
at higher pAg values, an excess of halide ions. Silver halide precipitations are normally carried out under excess halide ion conditions
(i.e., at high pAg, low pX).

Silver halide can form water-soluble complexes under conditions
in which excess halide ions or silver ions are present. Under excess
halide ion conditions, this complex ion formation can be represented
by

$$[Ag^+] + n[X^-] \rightleftharpoons [AgX_n^{1-n}] \tag{4}$$

Therefore, the total solubility (total soluble silver in the solution
phase) of AgX crystals in equilibrium with excess halide ions should
include both the free silver ion and the complex ions:

$$C_s = [Ag^+] + [AgX] + [AgX_2^-] + [AgX_3^{2-}] + \cdots$$

$$= K_{sp} \left\{ \frac{1}{[X^-]} + \beta_1 + \beta_2[X^-] + \beta_3[X^-]^2 + \cdots \right\} \tag{5}$$

where β_n is the stability constant for the nth complex ion. Figure 1
shows the calculated total solubility of AgBr versus pAg at different
temperatures using reported β_n values and their associated enthalpies
[1]. This plot represents solubility behavior on the excess halide
side of the equivalence point. Solubility increases as temperature
is increased. The shape of the curves can be understood in terms of
the dissociation reaction and the complex ion equilibria. For a
slight excess of halide ions, the solubility is reduced compared to
that at the equivalence point, due to the common ion effect. However,

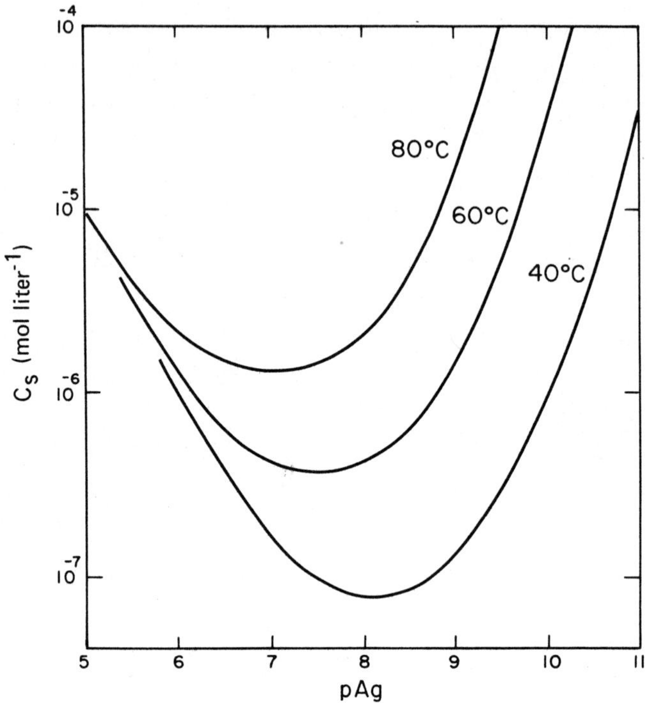

FIG. 1. Total solubility C_S of AgBr as a function of temperature and pAg (= $-\log[Ag^+]$).

at a higher excess of halide ions, the solubility increases because complex ion formation outweighs the common ion effect.

The solubility of AgX also depends on the specific halide present in the crystal and on the influence of AgX complexing agents. Generally, solubility is increased by the presence of a complexing agent and by a change in halide type from iodide to bromide to chloride. The effect of solubility on the crystallization processes in AgX precipitations will be discussed later.

B. Basic Precipitation Schemes

In general there are two common methods of AgX precipitation: single-jet and double-jet [1-3]. In the single-jet method, the silver nitrate

solution is introduced at a carefully controlled rate over a period
of time into a stirred solution containing the halide salt and gelatin. In this instance, immediately as the precipitation begins, a
large excess of halide ions (high pAg) is present in the reaction
vessel. This excess halide ion concentration gradually decreases.
The pAg may drop by 1 to 3 units over the course of the precipitation.

In the double-jet method the silver nitrate solution and the
halide salt solution are added simultaneously through separate input
lines to a stirred gelatin solution. The addition rates of the two
reactant solutions are generally balanced to maintain a desired pAg
in the precipitation vessel. The control of pAg can be carried out
continuously and automatically with the feedback control system described by Claes and Berendsen [6a]. The effect of pAg (or pBr) on
the final shape of AgBr crystals in double-jet precipitations will be
discussed later.

In addition to these batch operations, AgX precipitation can
also be carried out by a continuous method [7,8]. Continuous streams
of reactants (silver nitrate and halide salt) and gelatin are fed to
a stirred vessel while product is simultaneously removed to maintain
a constant reaction volume. The pAg control system used in double-jet
precipitations can also be applied here. Following a transient time
period, a steady state is reached, after which the size distribution
and shape of the AgX crystals removed from the precipitation vessel
remain unchanged. Figure 2 shows schematic diagrams for the three
basic precipitation methods.

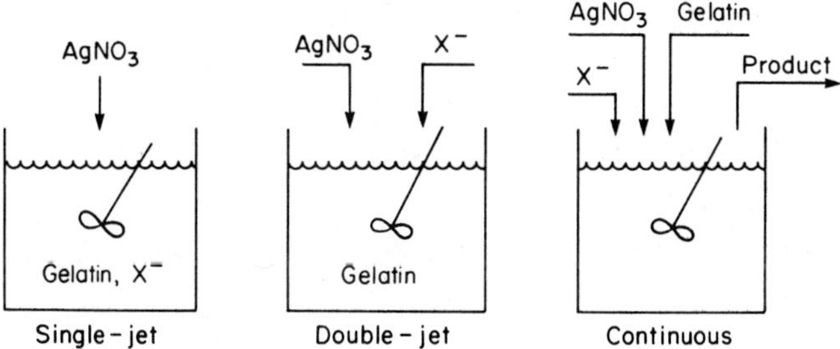

FIG. 2. Schematic diagrams of basic AgX precipitation methods.

The shape and size distribution of the AgX crystals are influenced by the choice of precipitation scheme. For example, the double-jet method can produce crystals with uniform size and shape, whereas the single-jet method normally gives a much wider distribution of crystals with different shapes. Under certain conditions, the continuous-precipitation method yields crystals with uniform shape and polydisperse size distributions [8]. Typical electron micrographs of AgBr crystals obtained from the three precipitation schemes are shown in Fig. 3.

The precipitation of AgX crystals is affected by several variables other than the precipitation scheme. These include: rate of reactant addition, degree of agitation, precipitation temperature, pAg (or pX), pH, halide content, peptizer type and concentration, and the presence of an AgX complexing agent or growth restrainer, etc. The time of precipitation (run time) is important for single-jet and double-jet precipitations. For the continuous scheme, the crystal size distribution is influenced by the average retention time in the reaction vessel and the suspension density of the AgX crystals [7-9]. Specific examples of the effects of these precipitation variables are given elsewhere [1-8].

C. Crystallization Processes

Silver halide precipitation consists of some or all of the following processes: nucleation, growth, Ostwald ripening, coalescence, and recrystallization [1,4]. *Nucleation* is the process by which new crystals of minute size are created. *Growth* is the addition of new material to existing crystals. *Ostwald ripening* is the process by which large crystals grow at the expense of small ones due to their difference in solubility (i.e., the small crystals dissolve into the solution, and the dissolved material subsequently precipitates onto the large crystals). *Coalescence* refers to an abrupt change in size when larger aggregates are formed by direct contact and welding together of crystals that were once separated. *Recrystallization* is the process in which the crystal composition changes due to the forma-

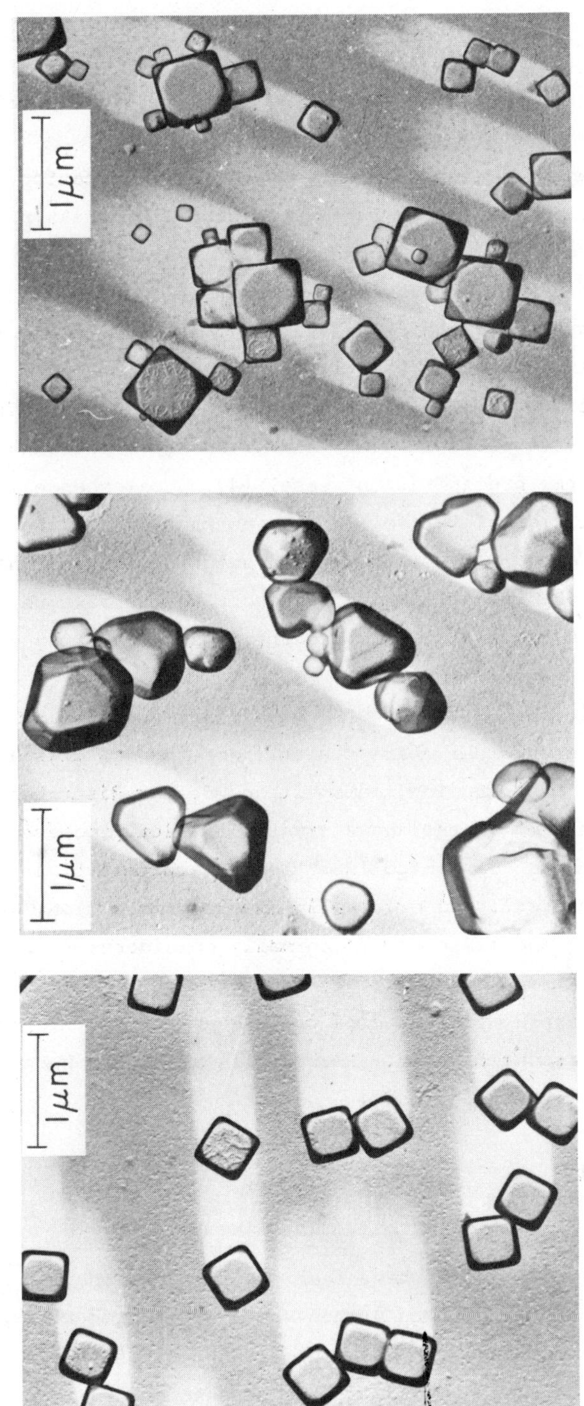

FIG. 3. Typical electron micrographs of AgBr crystals produced by the basic precipitation methods.

tion of solid solutions in a mixed AgX system (e.g., coprecipitation
of AgBr and AgI crystals).

All of these processes can occur during precipitation. Their
relative importance varies with the conditions. For example, under
normal double-jet conditions, observable nucleation is over very
rapidly (within 1 min) and further addition of reactant material results only in growth of the relatively monodisperse (uniform size)
AgX crystals already present [10,11]. Ostwald ripening of existing
crystals prevails when a high excess halide ion concentration (e.g.,
during a single-jet precipitation) and/or an AgX complexing agent
are present in the precipitation vessel [4]. Coalescence is important
in the precipitation of AgX sols in the absence of gelatin [12] or
when less than about 0.5 wt% of gelatin solution is used in the precipitation [13]. Recrystallization can occur when a mixture of silver
halides (AgCl, AgBr, and AgI) is present, thus permitting solid solutions to be formed [4].

The photographic response of a silver halide emulsion is largely
dependent on the physical characteristics of the AgX crystals (average size, size distribution, composition, defect distribution, etc.).
A good understanding of the above crystallization processes can provide a better basis for controlling these characteristics. The pur-
pose of this chapter is to critically examine nucleation, growth, and
Ostwald ripening. Coalescence is normally considered to be unimportant
in AgX precipitations because of the presence of a protective colloid
[1,4]. The possible occurrence of coalescence during the nucleation
stage will be discussed later. Recrystallization has been discussed
in great detail in Refs. 1 and 4.

II. NUCLEATION

A. Mechanisms of Nucleation

Several possible mechanisms have been suggested as the cause of crystal nucleation from a supersaturated solution. Three recognized
nucleation processes are (1) homogeneous nucleation, which occurs
spontaneously and does not require the presence of surfaces of any

kind; (2) heterogeneous nucleation, usually prompted by the presence of a foreign solid phase surface; (3) secondary nucleation, prompted by the presence of a crystalline phase of the solute material.

1. Homogeneous Nucleation

The classical theory of homogeneous nucleation treats the work of formation of the nucleus as the activation energy in a kinetic expression for the nucleation rate [14]. The size of the nucleus is obtained from the Gibbs-Thomson (Kelvin) equation and consequently is a function of supersaturation. The rate of nucleation increases very rapidly with increasing supersaturation, so that the formation of nuclei appears to begin at a critical supersaturation.

A more empirical approach to homogeneous nucleation was developed by Christiansen and Nielsen [15], who proposed a relationship between the induction period (the time interval between mixing two reacting solutions and the appearance of the precipitate) and the initial concentration of the supersaturated solution. The induction period represents the time needed for the formation of a nucleus and may range from microseconds to days depending on the supersaturation.

The classical theory and the empirical theory of Christiansen and Nielsen both use a clustering mechanism of the reacting ions. However, they do not agree on the effect of supersaturation on the size of a nucleus. The former theory predicts that the nucleus size depends on supersaturation, while the latter theory indicates a smaller but constant nucleus size. So far these differences have not been resolved, mainly because of the uncertainties associated with the experimental investigation of true homogeneous nucleation.

2. Heterogeneous Nucleation

Most systems of practical interest always contain tiny particles and nucleate at much lower supersaturations than required for homogeneous nucleation. The phenomenon of heterogeneous nucleation is thought to be the result of a foreign surface lowering the free energy change required for the formation of a nucleus [16]. At high supersaturations, homogeneous nucleation predominates over heterogeneous nucleation because its rate increases substantially and all of the foreign surface sites are exhausted. For precipitation of many kinds of sparingly soluble materials, Nielsen [17] found that heterogeneous nucleation occurred when fewer than 10^6 nuclei formed in 1 ml of solution. Homogeneous nucleation occurred when more than 10^6 nuclei formed per ml. More thorough discussion of homogeneous nucleation and heterogeneous nucleation is available in Refs. 18-20.

3. *Secondary Nucleation*

Secondary nucleation refers to the generation of new nuclei (secondary nuclei) in a supersaturated solution in which parent crystals of the solute material already exist. This phenomenon is frequently encountered in the chemical industry in suspension crystallization operations which involve slurries of crystals in agitated supersaturated solutions. The vast crystal surface area existing at any one time generally prevents the buildup of supersaturation to the level at which homogeneous or heterogeneous nucleation would occur.

Postulated mechanisms of secondary nucleation include crystal breakage, attrition, dendrite shearing, cluster shearing, and classical heterogeneous nucleation. The generation of secondary nuclei is believed to be most often due to breakage of a part of the parent crystal or removal of embryonic clusters of the solute material from the surface of the growing parent crystal. Fluid shear and mechanical contact (contacts between crystals or between a crystal and another solid subject) are thought to be responsible for removing the secondary nuclei from the solution-crystal interface. Substantial experimental work on this phenomenon has been done over the past decade. Generally, the results suggest that secondary nucleation is strongly influenced by the environmental conditions of crystallization, e.g., level of supersaturation, intensity of agitation, and mass of parent crystals in suspension. Thorough reviews of secondary nucleation phenomena were recently made by Estrin in Volume 2 of this series [21] and by Botsaris [21a].

B. Evidence of Homogeneous Nucleation

The traditional method of examining nucleation alone in a precipitation system has usually involved very slow addition of reactant solutions so that the nucleation does not occur for about an hour, and
subsequent crystal growth is greatly minimized. A common technique
for studying this nucleation is to monitor the disappearance of reactant ions from solution through changes in the conductivity of the
solution [22,23]. Davies and Jones [22] applied this technique to
study the precipitation of binderless AgCl sols by mixing solute solutions of silver nitrate and potassium chloride. They found that at
25°C the value of the concentration product for silver and chloride
ions at which nucleation occurred was about 1.7 to 2.0 times the solubility product, corresponding to a supersaturation ratio of 1.3 to 1.4.
(Supersaturation ratio is defined as the ratio of the bulk solute concentration to the solubility [24].) Klein et al. [23] conductimetric-
ally studied the precipitation of AgCl in binderless systems where the
chloride ion was slowly produced from hydrolysis of allyl chloride.
Nucleation did not begin until the supersaturation ratio reached a
value of about 1.7. By combining the experimental results with theoretical considerations, the size of a nucleus in AgCl was deduced to
be five ions [23]. However, such experiments, in which the reactants
are added very slowly, must be carried out in exceptionally clean
systems. Otherwise, nucleation is almost surely influenced by the
heterogeneous process.

The rates of reactant addition in practical AgX precipitations
are several orders of magnitude higher than those in the above experiments. Also, the starting materials ($AgNO_3$, halide salt, and gelatin)
are not normally made scrupulously clean. Thus, extrapolation of results from the above studies to practical AgX systems is difficult to
make. However, in a typical double-jet precipitation of AgBr at 50°C,
Berry and Skillman [10] observed that the number of crystals present
was about 10^{12} ml^{-1} at the end of 1 sec and 4×10^{13} ml^{-1} at the end

of the precipitation. These values are well above the region of 10^6 crystals ml^{-1} where Nielsen [17] concluded that homogeneous nucleation occurred in the precipitation of several sparingly soluble materials.

Berry and Skillman [4] attempted to change the number of AgX crystals generated during precipitation by adding various concentrations of foreign particles to the precipitation vessel. CuS particles of 25 Å and AgI crystals of 150 Å, at concentrations of about 10^{15} and 10^{12} ml^{-1}, respectively, were used as potential heterogeneous nucleation sites in a series of AgBr double-jet precipitations. No significant change in the size or number of the resulting AgBr crystals was detected, within experimental uncertainty, when these foreign particles were present. Berry and Skillman concluded that heterogeneous nucleation is unimportant even in the presence of a high concentration of foreign particles and that homogeneous nucleation is dominant in the precipitation of AgX photographic emulsions [4].

The nucleation kinetics of AgBr crystals were studied by Wey et al. [8] using a continuous-suspension crystallizer operated under steady state conditions. Ths population balance technique [9] was applied to the analysis of crystal size-distribution data obtained under different precipitation conditions. They found that the nucleation rate depended on supersaturation to the fourth power and was insensitive to the level of agitation and AgBr suspension density [8]. Since secondary nucleation is a sensitive function of agitation and suspension density and, in general, has a relatively low supersaturation dependence [21,21a], their results gave no indication that a secondary nucleation mechanism is involved in silver halide precipitation. This conclusion is understandable if one considers the very minute size (<3 μm) of AgX crystals produced for photographic emulsions. The effect of fluid shear or mechanical contact on nucleation should be insignificant because the motion of the crystal relative to the solution and the inertial forces of the crystal are very small. Furthermore, the gelatin solution can serve as a protective colloid for the AgX crystals and further reduce the possibility of secondary nucleation.

As pointed out by Berry and Skillman [4], mixing efficiency is an uncertain factor in AgX precipitation. Because of the sparingly soluble nature of AgX (solubilities ranging from 10^{-5} to 10^{-8} mol/liter), it is physically impossible to instantly dilute the concentrated reactant solutions (>1 mol/liter) normally used. The inadequacies in mixing make temporary, localized, high-reactant concentrations in the reaction vessel unavoidable. The actual supersaturation in the region of introduction of the reactant solutions is typically 10^{5} to 10^{8} times the solubility. These high local supersaturations are well above the region where Nielsen [17] observed the occurrence of homogeneous nucleation. Berry [25] and Gutoff et al. [26] considered the AgX precipitation vessel to contain two regions: the well-mixed bulk of the vessel and a region of extreme supersaturation where the highly concentrated reactants are introduced. They suggested that "transient" nuclei of small size are formed continually in the high-supersaturation region all during precipitation. These transient nuclei are continuously fed into the bulk of the vessel where the bulk supersaturation determines their survival. For example, in the very early stage of a double-jet precipitation (within 1 min), the bulk supersaturation is relatively high and most of the transient nuclei survive in the bulk of the vessel (i.e., observable nucleation occurs). These surviving nuclei grow rapidly into crystals and provide enough surface area to reduce the bulk supersaturation below the level required for the survival of the continuously generated transient nuclei, thus ending the observable nucleation. From this point on, the newly generated transient nuclei dissolve by Ostwald ripening in the bulk solution, and act as a source of additional silver halide for the growing crystals. For single-jet precipitations, the generation and survival of the AgX nuclei are complicated by Ostwald ripening during the observable nucleation period. Because of Ostwald ripening, the observable nucleation may occur over a longer time than with double-jet precipitations.

Several attempts have been made to determine the nucleus size in practical AgX precipitations. Using electron microscopy and light

scattering (turbidity), Berry and Skillman [10] measured the sizes of
crystals formed during the first few seconds of a double-jet precipitation. They reported a cube edge length of 275 Å for AgBr crystals
at 50°C and pBr 3.2. Claes and Borginon [27] measured the sizes of
crystals in the initial stage of AgBr precipitation by a dye adsorption technique and found a crystal diameter of about 200 to 600 Å,
depending on the precipitation temperature. These measurements were,
however, made on crystals that had grown beyond the size of a nucleus.
Wey and Strong [28] examined the critical supersaturation condition
for nucleation in a seeded double-jet precipitation. They estimated
a nucleus size of 140 Å for AgBr at 70°C and pBr 3.3. These reported
values are much larger than those obtained from the classical treatment of homogeneous nucleation in solution. (A value of five ions
was reported for the nucleus size of AgCl in slow precipitation from
homogeneous solution [23].) Berry and Skillman [10] suggested that
the relatively large observed size, associated with rapid precipitation of AgX in a gelatin solution, resulted from Ostwald ripening in
the bulk solution. On the other hand, the growth of nuclei in the
high-supersaturation region (near the reactant introduction point)
can also contribute to the large observed size. In addition, the
coalescence of nuclei may offer another explanation. As suggested
by Ipatov [29], two processes could take place at the first moment
of AgX nucleation: coalescence of nuclei followed by formation of
the protective gelatin layer. That is, the gelatin layer may not
form at once on the nuclei. Thus, the coalescence of nuclei may occur
at this very moment, resulting in a large observed size. However, as
soon as the gelatin layer forms and is adsorbed onto the crystal surface, coalescence is greatly retarded and the crystals are mainly influenced by regular growth and Ostwald ripening.

C. Factors Affecting Nucleation

The formation of AgX precipitates is affected by several variables
such as the rate of reactant addition, solubility, and agitation condition. Wagner [30] derived the following equation for the total num-

ber of nuclei (per unit volume) formed during precipitation:

$$Z = \frac{RR_g T}{8\pi TDvC_s} \qquad (6)$$

Here R represents the rate of reactant addition, R_g is the gas constant, T is absolute temperature, T is the surface energy of the crystal, D is the diffusion coefficient of the solute in the solution, v
the molar volume, and C_s is the solubility of the precipitate. Equation (6) suggests that the number of nuclei formed is directly proportional to the rate of reactant addition and inversely proportional
to the solute solubility.

In an AgBr double-jet precipitation, Klein and Moisar [31] confirmed the direct proportionality between the number of nuclei (Z)
and the rate of reactant addition (R) when R values were less than
1.7×10^{-8} mol sec^{-1} ml^{-1}. However, at higher R values Z was essentially independent of R, as shown in Fig. 4. To explain these experiment-
al observations, they suggested that the number of nuclei (Z) predicted by Eq. (6) could be different from the number of *initial* nuclei
(Z_0) generated at the very beginning of the precipitation. At low R

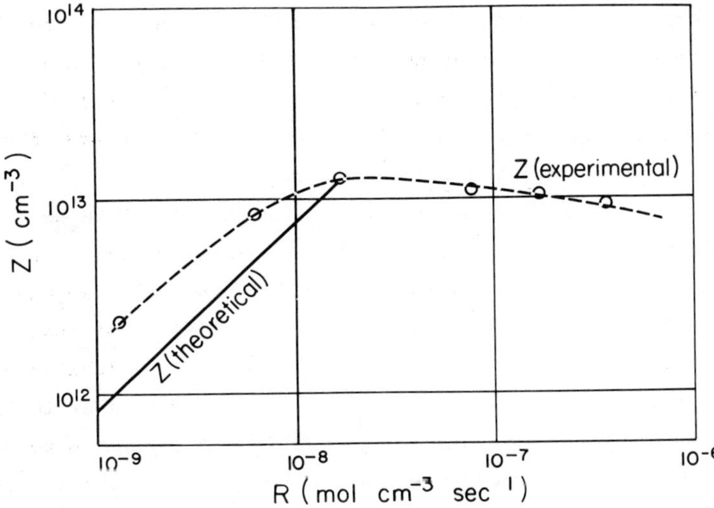

FIG. 4. Total number of nuclei per unit volume Z as a function of
the rate of reactant addition R. (From Ref. 31; reprinted with
permission.)

values, dissolution of initial nuclei may take place so that Z is smaller than Z_0. In this case, an increase in R would increase the survival of initial nuclei (i.e., Z is increased) so that a direct relationship between Z and R might be expected. However, according to Klein and Moisar, when R reaches a critical value at which Z is equal to Z_0, all the initial nuclei survive and further increases in R should have no effect on Z.

Loginov and Denisova [32] studied the effect of reactant addition rate on the number of nuclei formed during double-jet precipitation. At 42°C and at pAg 6.4 for AgI and 9.0 for AgBr, they again confirmed the direct proportionality between Z and R for both AgI and AgBr crystals. However, when R was greater than 3×10^{-7} mol sec^{-1} ml^{-1}, Z decreased sharply. The decrease in Z was attributed to the coalescence of the large number of initial nuclei generated at high R values [32]. Note that the critical R value here is about one order of magnitude higher than that reported by Klein and Moisar [31]. The relationship between the number of nuclei and the rate of reactant addition under different AgX precipitation conditions should be further explored.

According to Eq. (6), the number of nuclei formed during precipitation is inversely proportional to the solubility. Thus, as the solubility is increased the number of nuclei decreases, while from a mass balance the crystal size must increase. Berry and Skillman [10] observed that lowering the temperature from 50 to 40°C in an AgBr double-jet precipitation decreased the average volume of a crystal by a factor of 2, due to the decrease in solubility. Adding 0.2-N NH_4OH to the vessel before precipitation increased the average volume of a crystal by about 200 times, due to the solubility increase (NH_4OH is a common AgX complexing agent). Klein and Moisar [31] found that the number of nuclei decreased by a factor of nine when the solubility was increased 10 times (by changing from the less soluble AgBr to the more soluble AgCl). Their results agree well with Eq. (6). However, detailed correlations between the number of nuclei and solubility under different precipitation conditions (temperature, pAg, halide type, complexing agent, etc.) have not yet been established.

As indicated earlier, the formation of nuclei during AgX precipitation may depend on two processes: (1) generation of transient nuclei in the high-supersaturation region (near the reactant introduction point); and (2) survival of these transient nuclei in the bulk solution. In order to survive, the transient nuclei must be larger than the "critical size" (associated with the bulk supersaturation) determined by the Gibbs-Thomson equation. (This critical size represents the size above which a crystal or a nucleus grows spontaneously and below which it dissolves.) However, the size of transient nuclei may be influenced by the mixing conditions in the region of reactant introduction and by the interaction between the high-supersaturation region and the bulk solution. Moreover, agitation conditions in the bulk solution may be important since a nonuniform bulk supersaturation can result in different critical sizes for the survival of transient nuclei within the bulk of the precipitation vessel. Note that although agitation is one of the important variables in AgX precipitation, it has received scant attention [33] in the published literature.

The initial AgX precipitates produced during the nucleation period contain a high level of structural imperfections which are probably quenched in during the rapid initial nucleation [34]. Such imperfections were first deduced from the observation that fresh AgX crystals underwent much more rapid Ostwald ripening than the same crystals aged for a few minutes [35]. X-ray diffraction investigation of these microcrystals showed a high degree of structural disorder at an age of 2 sec after nucleation, which was largely gone after 5 min of aging [34].

III. CRYSTAL GROWTH

A. Theories of Crystal Growth from Solution

As soon as nuclei are formed in a supersaturated solution they begin to grow into crystals. Several theories have been proposed to explain the mechanism and rate of crystal growth. Reviews by Ohara and Reid [36] and Garside [37] detail recent advances in the kinetics of crystal growth from solution. Generally, growth theories may be

classified under three categories: (1) two-dimensional nucleation;
(2) screw dislocation; and (3) bulk diffusion.

1.   *Two-dimensional Nucleation Theories*

According to the two-dimensional nucleation growth concept, planar
crystal faces are energetically unfavorable for the formation of new
layers. To initiate crystal growth on a perfectly flat surface, a
two-dimensional nucleus must be generated to provide incorporation
sites (steps) for the growth units (ions, molecules, atoms). Several
models have been proposed for two-dimensional nucleation growth. Basically, they differ in the assumptions of the rate at which the growth
units spread across the surface compared with the time taken to form
a new nucleus, and in the number of growth units needed to form a
nucleus for a new layer [36].

In the mononuclear model, the limiting step is the formation of
a nucleus. Once one is formed, subsequent growth spreading across
the surface is infinitely rapid. Therefore, the linear growth rate,
G, may be given as [36,37]

$$G \propto L^2 \sigma^{1/2} \exp\left(\frac{-B}{\sigma}\right) \tag{7}$$

where L is the size (edge length or diameter) of the crystal, σ is the
supersaturation parameter defined by $(C_i - C_e)/C_e$, C_i is the interface
solute concentration, C_e is the solute concentration in equilibrium
with the solid phase, and B is a parameter. Based on Eq. (7), the
mononuclear model predicts the growth rate to be proportional to the
crystal surface area. For a polynuclear model, the spreading velocity
is taken as zero and the crystal surface can only be covered by the
accumulation of a sufficient number of nuclei. The growth rate can
be represented by [36,37]

$$G \propto \sigma^{-3/2} \exp\left(\frac{-B}{\sigma}\right) \tag{8}$$

Note that G is independent of crystal area (or size), and a maximum in
G would be expected at some value of σ. The above growth models represent two extreme cases. A third model, known as the "birth-and-spread"

model, allows for formation of nuclei and their subsequent growth at a finite rate. In this case, new nuclei can form on top of uncompleted layers, and the growth rate can be given as [36,37]

$$G \propto \sigma^{5/6} \exp\left(\frac{-B}{\sigma}\right) \tag{9}$$

This model predicts that G does not depend on crystal area, and no maximum of G with σ would be obtained.

These three models are sometimes unsatisfactory in predicting or correlating crystal growth rates. For example, at low values of σ, the predicted growth rates are often much less than those observed experimentally [24,36].

2. Screw Dislocation Theories

In order to explain the discrepancy between experimental observations and the two-dimensional nucleation theories, Frank [38] proposed a growth model based on dislocation theory. A screw dislocation can give rise to a spiral step which never disappears during growth. Hence, the formation of a two-dimensional nucleus at the crystal surface is not necessary to sustain growth. Burton, Cabrera, and Frank (BCF) [39] considered growth of the spiral to be limited by diffusion of adsorbed growth units across the crystal surface and developed a growth equation of the form

$$G \propto \frac{\sigma^2}{\sigma_c} \tanh\left(\frac{\sigma_c}{\sigma}\right) \tag{10}$$

where σ_c is a complex temperature-dependent constant which includes parameters depending on step spacings. At low supersaturations, the BCF equation approximates to $G \propto \sigma^2$, and at high supersaturations to $G \propto \sigma$.

A simple power-law equation of the form

$$G = K_i (C_i - C_e)^m \tag{11}$$

has been used frequently to represent the processes occurring at the crystal surface. This clearly represents the two limiting cases of the BCF equation and is also a good approximation to the two-dimensional

nucleation models over a limited range of supersaturation [37]. The
above-mentioned nucleation and dislocation growth models are sometimes
included under the general title of "surface reaction" or "surface
integration" growth. Note that the surface integration constant K_i
is independent of crystal size except for the mononuclear growth model
[Eq. (7)] where $K_i \propto L^2$.

3. *Bulk Diffusion Theories*

The growth of crystals from solution is generally considered to involve two processes occurring sequentially: the mass transport of
growth units from the bulk supersaturated solution to the crystal-
solution interface by bulk diffusion and the incorporation of growth
units into the crystal lattice by surface integration. The models
of surface integration growth described earlier are all based on the
assumption that the resistance offered by bulk diffusion is negligible
compared to that offered by surface integration. However, if the surface integration rate is very rapid, the rate-limiting step of crystal
growth will be bulk diffusion.

The simplest model of bulk diffusion is the molecular diffusion
of growth units through a (fictitious) stagnant diffusion layer adjacent to the crystal surface with complete mixing beyond. For a diffusion layer thickness of δ, the linear growth rate can be derived by
integrating Fick's law and is given by the equation

$$G = \frac{4D}{\rho_c} \left[\frac{1}{L} + \frac{1}{2\delta} \right] (C - C_i) \tag{12}$$

where ρ_c is the crystal density, L is the crystal size (edge length
or diameter), and C is the bulk solute concentration. Although Eq.
(12) was derived primarily for a spherical crystal, it may also, to
the first approximation, be applied to other crystal shapes such as
cubic, octahedral, etc. A more thorough treatment of bulk diffusion
about a growing cube was recently done by Wilcox [39a]. When the
diffusion layer thickness is small compared to the crystal size (2δ
$\ll$ L), Eq. (12) can be reduced to

$$G = \frac{2D}{\rho_c \delta} (C - C_i) \tag{13}$$

Equation (13) suggests that G is inversely proportional to δ and is independent of L. However, when the diffusion layer thickness is large compared to the crystal size ($2\delta \gg L$), Eq. (12) becomes

$$G = \frac{4D}{\rho_c L} (C - C_i) \tag{14}$$

Here, G is independent of δ but is inversely proportional to L. Note that Eq. (14) also represents molecular diffusion in an infinite unstirred solution.

One of the difficulties associated with the stagnant diffusion layer model is the uncertainty in determining δ. This leads one to consider the semiempirical engineering model

$$G = \frac{2k_d}{\rho_c} (C - C_i) \tag{15}$$

Here, k_d is the mass transfer coefficient [40] which accounts for the fluid dynamics and the physical properties important to the process. For mass transfer associated with suspended crystals in an agitated solution, bulk diffusion ordinarily involves both molecular diffusion and convective transport. Experimental evidence [41] indicates that k_d can be reasonably well correlated by the Frössling equation [42] for spheres, cubes, and most other simple geometric shapes:

$$Sh = 2 + 1.10Re^{1/2}Sc^{1/3} \tag{16}$$

where $Sh = k_d L/D$ is the Sherwood number, $Re = u_s L\rho/\mu$ is the Reynolds number, u_s is the particle "slip" velocity, ρ is the solution density, μ is the solution viscosity, and $Sc = \mu/\rho D$ is the Schmidt number.

The first term on the right-hand side of Eq. (16) represents the contribution of molecular diffusion. The second term is for convective transport. At one extreme, if the solid particles are very small (e.g., typical AgX crystals), the suspended phase tends to move with no slip along with the circulating fluid (i.e., $u_s = 0$). In this case, the effect of convective transport is negligible and bulk diffusion occurs mainly by molecular diffusion. Thus,

$$Sh = k_d \frac{L}{D} = 2 \tag{17}$$

Substitution of Eq. (17) into Eq. (15) gives Eq. (14), as expected.

Heat of crystallization evolved at the crystal-solution interface may cause the temperature at this point to differ from that in the bulk solution. The supersaturation for growth will thus be different from that calculated on the basis of the equilibrium concentration at the bulk temperature. Ohara and Reid [36] considered the magnitude of this thermal effect and found it to be unimportant for crystal growth from solution.

B. Growth of Silver Halides

In double-jet precipitation, the number of AgX crystals quickly reaches a constant value. Further addition of reactants causes only growth of the relatively monodisperse crystals already present. In this case, growth rates of crystals can be determined from a simple mass balance or from measurement of electron micrographs of crystal samples obtained at successive times during the precipitation. If the level of solution supersaturation could be measured, experimental growth rates could then be correlated with the above-mentioned theoretical models to elucidate possible growth mechanisms. Unfortunately, for the AgX system the supersaturation is generally too small to be measured accurately by existing techniques. Nevertheless, the growth rate behavior can still be inferred by observing the size distribution or by following the relative sizes of two populations of crystals during double-jet precipitation [11,43-47].

Berry and Skillman [43,44] examined the size distribution of AgX crystals during the growth stage of a balanced double-jet precipitation. They observed a virtually constant width of the size distribution for the growth of well-formed AgBr and AgCl crystals. They found a size-independent growth behavior in accordance with both the two-dimensional nucleation models [Eqs. (8) and (9)] and the BCF theory [Eq. (10)]. Berriman [11], on the other hand, observed a substantial widening of the size distribution during an AgBr double-jet precipitation

His results suggested that small crystals grew at a slower rate than large ones in the same growth environment. Berriman [11] explained the distribution widening on the basis of the classical equation relating solubility and crystal size (the Gibbs-Thomson effect [24,48]). Berry [4] suggested that mild Ostwald ripening might also have occurred in Berriman's experiments because of the presence of some rounded crystals which would normally appear during Ostwald ripening of cubic AgBr crystals.

Hirata and Hohnishi [45] measured the size change for two distinct populations (0.21 and 0.43 μm) of monodisperse AgBr cubic crystals growing together in a common double-jet environment. They found that the small crystals grew more rapidly than the large ones. This size-dependent growth behavior was explained by a molecular diffusion model with a stagnant diffusion layer [Eq. (12)]. The thickness δ of this diffusion layer was estimated to be of the same order as the crystal size [45].

Wey and Strong [28] presented an experimental technique to examine the relationship between crystal size and growth rate for AgBr cubic crystals at a critical supersaturation condition at which nucleation just begins to occur. Their approach involved balanced double-jet precipitation at constant temperature ($70°C$) and bromide ion concentration (pBr 3.3) in suspension. For each experiment, the reaction vessel was charged initially with monodisperse AgBr seed crystals. By decreasing the number of seed crystals or increasing the reactant addition rate from one experiment to the next, nucleation eventually occurred when the system reached the critical supersaturation point. The corresponding critical growth rate was determined from a simple mass balance. The same experiments were repeated for five different monodisperse AgBr seed sizes ranging from 0.047 to 0.64 μm. Since these growth rates for different seed sizes were obtained at the same critical supersaturation, the effect of supersaturation can be eliminated from the correlation between growth rate and crystal size. The growth rate behavior is shown in Fig. 5 where the growth rate at critical supersaturation is plotted against crystal size. It is

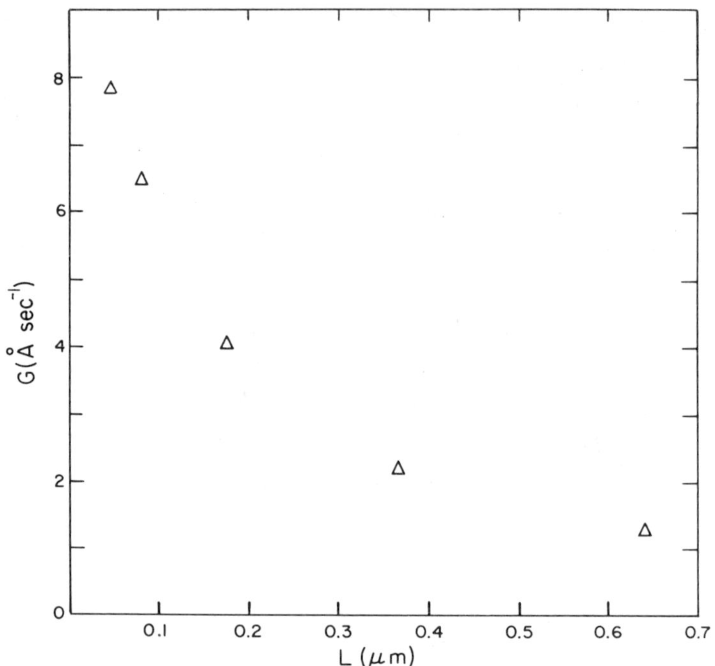

FIG. 5. Critical growth rate G of AgBr as a function of crystal size
L (constant rate of reactant addition for all runs). (From Ref. 28;
reprinted with permission.)

interesting that the small crystals grew faster than the large ones.
Although this size-dependent behavior is qualitatively similar to that
reported by Hirata and Hohnishi [45], the stagnant diffusion layer
model [Eq. (12)] cannot fully explain the observed growth behavior
since a plot of G versus 1/L for the experimental data does not give
a straight line as predicted by the model [28].

To explain the above size-dependent behavior for AgBr, Wey and
Strong [28] considered a growth model based on molecular diffusion
[Eq. (14)] combined with first-order surface integration [Eq. (11)
with m = 1]. The effect due to convective transport can be estimated
from the second term in the Frössling equation [42]. The Reynolds
number in the Frössling equation includes the particle "slip" velocity
which may be determined from the empirical equation suggested by Miller
[41]:

$$\frac{u_s}{u_t} = 6.44 \times 10^{-4} \, (\text{rpm})^{1.239} \tag{18}$$

where u_t is the terminal velocity of the particle which can be obtained from Stokes' law,

$$u_t = \frac{g_c L^2 (\rho_c - \rho)}{18\mu} \tag{19}$$

For 1-μm AgBr crystals suspended in an agitated 3 wt% gelatin solution, the following data may be used: $L = 10^{-4}$ cm, $\rho_c = 6.47$ g cm^{-3}, $\rho = 1$ g cm^{-3}, $\mu = 0.02$ g cm^{-1} sec^{-1}, $D = 2 \times 10^{-5}$ cm^2 sec^{-1}, $g_c = 980$ cm sec^{-2}, rpm = 2000. Substituting these values into the above equations gives $u_s = 1.2 \times 10^{-3}$ cm sec^{-1}, Re = 6×10^{-6}, Sc = 10^3. Therefore, Eq. (17) becomes

$$\begin{aligned}
\text{Sh} &= 2 + 1.10(6 \times 10^{-6})^{1/2} \, (10^3)^{1/3} \\
&= 2 + 0.03 \approx 2
\end{aligned} \tag{20}$$

Equation (20) indicates that for submicron AgX crystals, the effect due to convective transport is less than 2% of the overall bulk diffusion. Thus the use of the molecular diffusion model [Eq. (14)] in the AgX system is justified.

The overall growth rate expression based on molecular diffusion and first-order surface integration can be given by

$$G = \frac{K_i (C - C_e)}{1 + \varepsilon L} \tag{21}$$

where

$$\varepsilon = \frac{\rho_c K_i}{4D} \tag{22}$$

The experimental data of Wey and Strong [28] correlate very well with Eq. (21), as shown in the plot of 1/G versus L in Fig. 6. From the slope and intercept of this plot, the value of ε can be estimated. It is interesting that an ε value of about 16 μm^{-1} was obtained for AgBr growth at 70°C and pBr 3.3. Since ε represents the ratio of the "resistance" of molecular diffusion to that of surface integration,

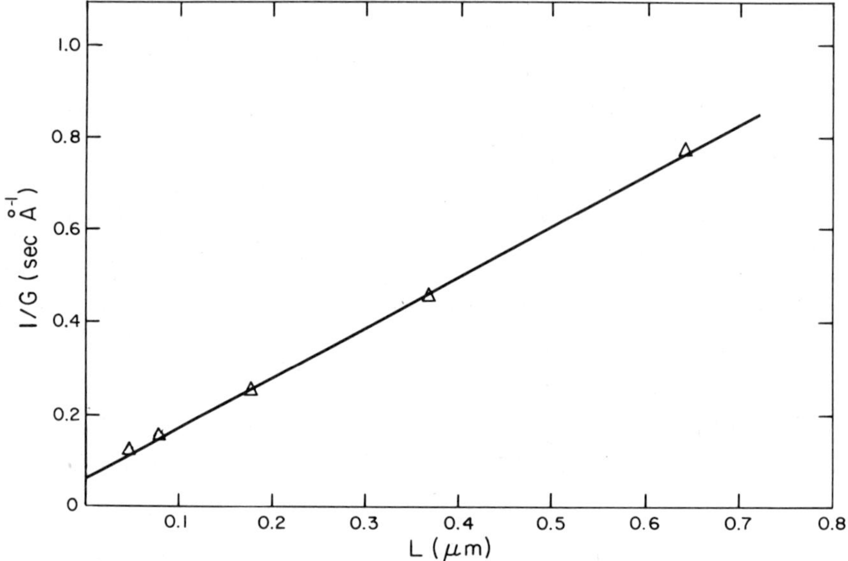

FIG. 6.   Reciprocal of critical growth rate 1/G of AgBr as a function of crystal size L.   (From Ref. 28; reprinted with permission.)

the results suggest that molecular diffusion is important in most conventional AgBr precipitations [28]. It is worth mentioning that the initial seeding technique, proposed by Wey and Strong [28] for studying AgBr growth behavior, was recently applied to the AgCl system [47]. At 60°C and pCl 2.15, the AgCl crystals showed a size-dependent growth behavior similar to that for AgBr crystals. However, the growth of AgCl crystals was found to be dominated entirely by molecular diffusion [47]. The activation energy for AgCl crystal growth was estimated to be about 4.9 kcal mol^{-1} at pCl 2.15 over a temperature range of 40 to 80°C [47].

The influence of the Gibbs-Thomson effect [24,48] was considered by Wey and Strong [46] to explain the various contradictory growth rate behaviors reported in the literature for AgBr [11,28,43,45]. Rearranging the Gibbs-Thomson equation, we obtain the following relation between equilibrium solute concentration and crystal size,

$$C_e(L) = C_s \exp\left(\frac{\Gamma_D}{L}\right) \simeq C_s\left(1 + \frac{\Gamma_D}{L}\right) \tag{23}$$

where

$$\Gamma_D = \frac{4vT}{R_g Tv} \tag{24}$$

Here, the stoichiometric factor v is roughly equal to one for AgX under excess halide ion concentrations practically used [48a]. The solubility C_s is equivalent to the equilibrium solute concentration associated with a flat crystal surface (i.e., a very large crystal). Equation (23) indicates that the equilibrium solute concentration of small crystals is always greater than that of large ones. Therefore, in a common growth environment, the "effective" driving force for the growth of small crystals is smaller than that for large crystals. Substitution of Eq. (23) into Eq. (21) gives a growth rate expression which includes molecular diffusion, surface integration, and the Gibbs-Thomson effect,

$$G = \frac{g(1 - L^*/L)}{1 + \varepsilon L} \tag{25}$$

where

$$g = K_i(C - C_s) \tag{26}$$

$$L^* = \frac{\Gamma_D}{\ln S} \tag{27}$$

Here, L^* represents the critical size above which a crystal (or a nucleus) grows spontaneously and below which it dissolves. S, normally referred to as the "supersaturation ratio," is equal to C/C_s. The influence of the Gibbs-Thomson effect on AgBr crystal growth is illustrated in Fig. 7 by the plot of normalized growth rate G/g versus L for different L^* values using a value of 16 μm^{-1} for ε. The curve with $L^* = 0$ represents the case when the Gibbs-Thomson effect is not important [i.e., Eq. (25) is reduced to Eq. (21)]. This behavior was used to explain the influence of molecular diffusion and surface integration on the size dependency of AgBr crystals growing in a highly supersaturated environment [28]. As the Gibbs-Thomson effect becomes more important (reflected by larger L^*), the growth rates in the small size region exhibit greater deviation from the $L^* = 0$ curve.

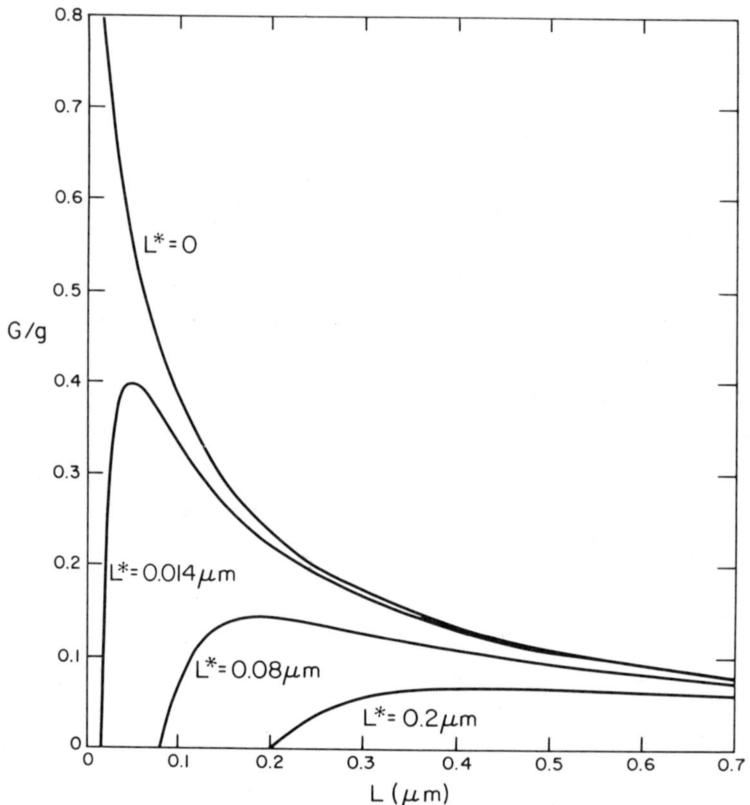

FIG. 7. Calculated (normalized) growth rate G/g as a function of crystal size L with critical size L^* as a parameter. (From Ref. 46; reprinted with permission.)

This decrease in growth rate is more pronounced when the crystal size is less than 2 to 3 times L^* [49,50]. Since L^* is related to the supersaturation ratio S [Eq. (27)], the size-dependent growth behavior may be altered by changing S. This can be done by varying the precipitation conditions thereby altering C and C_s.

To demonstrate the possible influence of the Gibbs-Thomson effect, Wey and Strong [46] examined the growth behavior of two distinct populations of AgBr crystals (0.08- and 0.64-μm cubes) growing together under various double-jet precipitation conditions. Both constant and

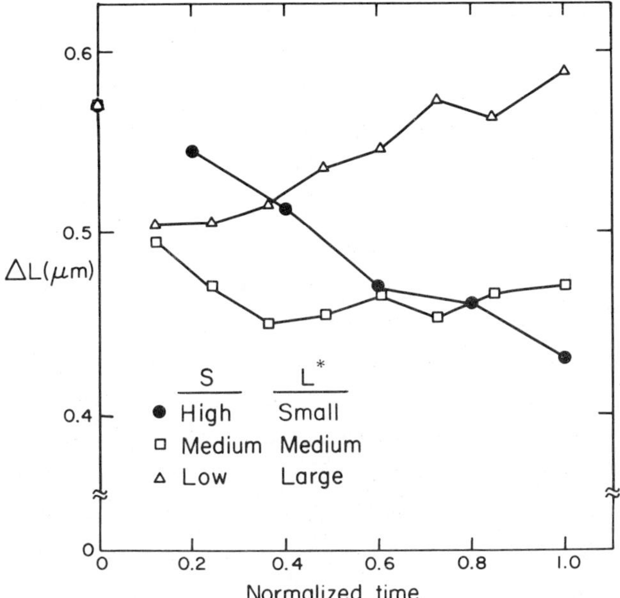

FIG. 8.   Difference in edge length between two populations of AgBr crystals during growth under different supersaturation conditions. (From Ref. 46; reprinted with permission.)

accelerated reactant addition schemes were used to achieve the desired variation in C. For the latter, the rate of reactant addition was gradually increased throughout the run. The excess bromide-ion concentration in suspension was also varied to alter C_s. These variations in C and C_s caused changes in S and L^*. Figure 8 shows the difference in edge length between the two populations during growth under three different levels of the supersaturation ratio. These results are qualitatively consistent with the theoretical considerations. At a high S (small L^*), the growth of AgBr is dominated by molecular diffusion and surface integration so that small crystals grow faster than large ones. Conversely, at a low S (large L^*), the dominant influence on crystal growth is the Gibbs-Thomson effect, which causes a substantial reduction in the growth rate of small crystals relative to large ones. Therefore, the small AgBr crystals grow at a lower rate than the large ones. An intermediate S results in roughly counterbalanced influences of the Gibbs-Thomson effect and of molecular diffusion and surface integration (i.e., nearly size-independent growth can occur). Thus, the contradictory observations of AgBr growth behavior reported in the literature [11,28,43,45] can be qualitatively explained by considering the relative importance of the Gibbs-Thomson effect and of molecular diffusion and surface integration under different AgBr double-jet precipitation conditions.

The above discussion on the influence of the Gibbs-Thomson effect does not consider cases which involve extremely high halide ion concentrations (as found in single-jet precipitations) or the use of AgX complexing agents. Under these precipitation conditions, dissolution (or Ostwald ripening) of some of the existing crystals (especially the very small crystals) may occur even though new reactant material is continuously introduced [4,51,52]. Ostwald ripening of AgX crystals will be discussed later.

C. Factors Affecting Crystal Growth

The morphology (shape) of AgX crystals is influenced by the excess halide ion concentration in suspension during the crystal growing process. For example, at low bromide ion concentrations ($Br^- < 5 \times 10^{-4}$ mol liter^{-1} or pBr > 3.3), AgBr crystals grow in the form of cubes bounded by (100) faces. At higher bromide ion concentrations, the AgBr crystals grow as octahedra bounded by (111) faces. Figure 9 shows electron micrographs of typical AgBr cubes and octahedra. The structures of these two faces are quite different, as shown schematically in Fig. 10. The (100) faces consist of equal numbers of positive silver ions and negative bromide ions. They bear no excess electrical charge and may be considered as smooth planes. The (111) faces, on the other hand, contain only ions of the same charge with ions of opposite charge occupying the lattice sites of the layer below. Thus, an excess electrical charge can be created by the great number of noncompensated ions in the outermost crystal plane. Furthermore, a certain roughness is likely to occur on the (111) faces.

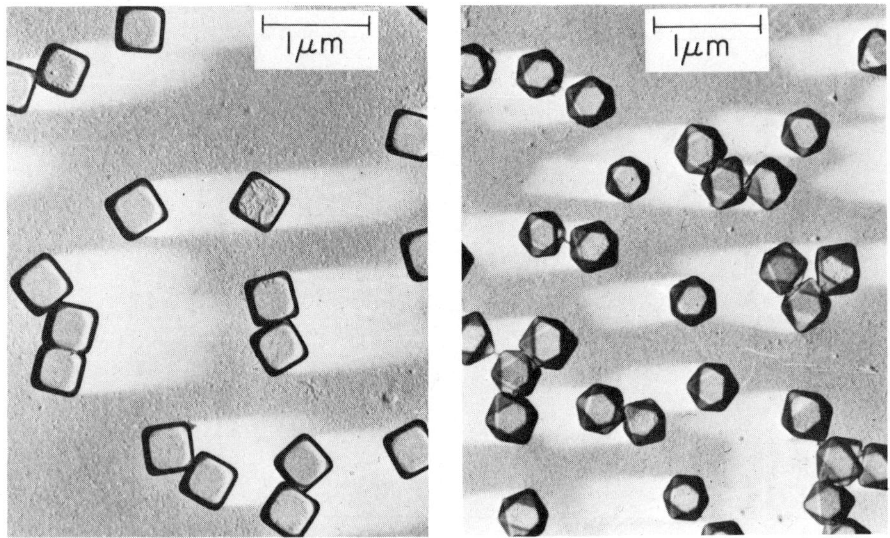

FIG. 9. Electron micrographs of typical AgBr cubic and octahedral crystals.

The change in morphology between (100) and (111) faces is determined by the relative growth rates of these two faces. Usually, the rapidly growing faces leave the growing process, and only faces with lower growth rates remain. To examine the change in relative growth rates of (100) and (111) faces as a function of excess bromide ion concentration, Moisar and Klein [53] measured the adsorptive property

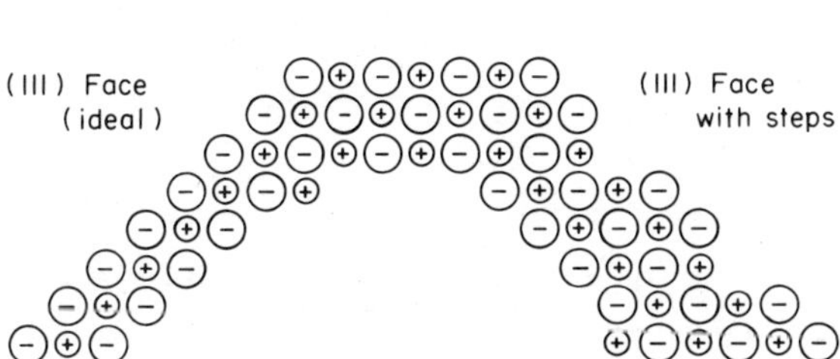

FIG. 10.   Arrangement of ions in an AgBr crystal.   (From Ref. 5; reprinted with permission.)

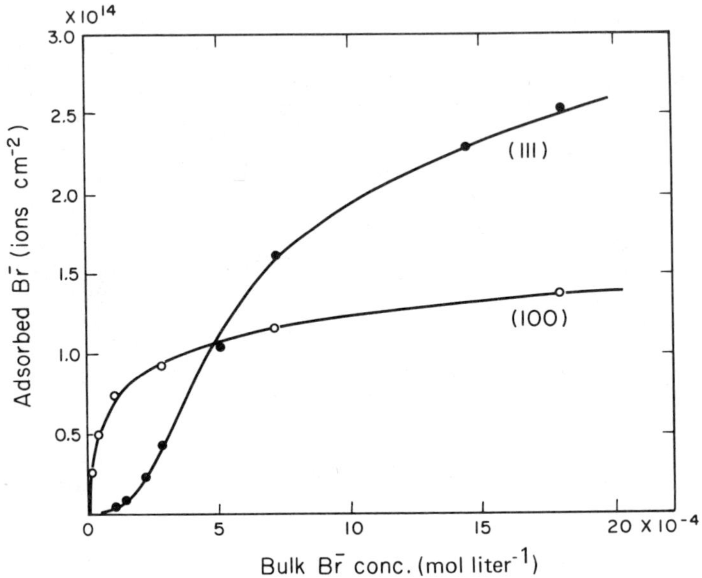

FIG. 11. Adsorption isotherms of bromide ion on AgBr (100) and (111)
faces. (From Ref. 53; reprinted with permission.)

of these two faces using a differential titration technique. Figure
11 shows the number of bromide ions adsorbed on (100) and (111) AgBr
faces versus the bromide ion concentration in the solution. The bromide adsorption on the (100) faces follows a Langmuir isotherm [54]
and increases rapidly to an asymptotic value. By contrast, the bromide adsorption on the (111) faces is much less at a low bromide ion
concentration in solution. However, at bromide ion concentrations
higher than 5×10^{-4} mol liter^{-1}, the (111) bromide adsorption rises
rapidly above the (100) asymptotic value [53]. (Recent work by Gardner et al. [54a] could not confirm the weak bromide adsorption on
the (111) faces at low bromide ion concentrations.) Moisar and Klein
[53] suggested from their data that at low concentrations of bromide
ions in solution, the (111) faces grow faster than the (100) faces
since fewer bromide ions are adsorbed on the (111) faces. Also, the
greater roughness of the (111) faces favors the deposition of solute
materials. This leads to the formation of cubic crystals. On the

other hand, primarily the (100) faces grow at high bromide ion concentrations because the excess bromide ions adsorbed on the (111)
faces retard the growth of the (111) faces due to an increase in
nucleation energy [53] and/or a repulsion of the negatively charged
complex ions which are the main source of silver ions under these
conditions [51]. This eventually leads to the formation of AgBr octahedral crystals. It is interesting to note that although AgCl crystals have the same ionic NaCl-type cubic lattice as AgBr crystals,
generally only the (100) faces have been reported even at very high
chloride ion concentrations. (Claes et al. [54b] have reported the
formation of (111) and (110) faces for AgCl crystals but these were
obtained in the presence of growth modifiers.)

The morphology of AgBr crystals is influenced by the presence of
the complexing agent NH_4OH, which is often used to increase the solubility of AgX crystals. In the presence of NH_4OH, cubic AgBr crystals
are obtained under high bromide ion concentrations which in the absence of NH_4OH would give octahedral crystals, i.e., the stability of
the (100) faces is shifted in NH_4OH solutions toward higher bromide
ion concentrations [53,55,56]. Moisar [5] found that bromide adsorption at the AgBr (111) faces decreased with increasing NH_4OH concentration. Therefore, the shape-determining influence of bromide ions
on the (111) faces is diminished at higher NH_4OH concentrations.
Markocki and Zaleski [56] illustrated the range in which the AgBr
(100) faces and (111) faces formed under different levels of NH_4OH
and excess bromide ion concentrations.

Small amounts of iodide ions (<12 mol%) are sometimes added to
an AgBr precipitation to form silver iodobromide solid solution crystals. The addition of iodide leads to an increase in the number of
crystals and a decrease in the average crystal size. The iodide content also has a substantial effect on the AgBr crystal habit [56–58].
Iodide favors the formation of the (111) faces, which become more
pronounced as the iodide content increases. Claes and Peelaers [57]
suggested that solvation of the crystal faces is strong for the (111)
faces which have high polarity, and weak for the (100) faces which

have low polarity. The direction of growth of a crystal depends on
the state of solvation of different faces. The incorporation of the
weakly solvated iodide ions can cause desolvation of the crystal surfaces [57]. This effect is most perceptible on the weakly solvated
(100) faces. Thus, the rate increase for the (100) faces is higher
than that for the (111) faces. This will favor the formation of the
(111) faces. Claes and Peelaers [59] also studied the effect of
strongly solvated cations such as Pb^{2+} or Cd^{2+} on crystal morphology.
They found that growth of AgBr in the presence of these impurity ions
favors the formation of the (100) faces.

During the growth of AgBr crystals, twinning can occur on the
(111) faces, especially with a high bromide ion excess. This is evi-
dent from the presence of crystal shapes which are different from
regular cubes or octahedra [43,60]. Berriman and Herz [61] examined
AgBr tabular hexagons and triangles with an x-ray microbeam and found
doubled sets of diffraction spots in the expected orientations for
(111) twinning. They suggested that these imperfect crystals were
formed by a twinning mechanism and not by screw dislocations. The
cause of twinning is unknown, although it was suggested [55] that
twinning may occur when a large complex ion of silver at a growing
surface interferes with the deposition of silver ions or bromide
ions in their usual positions. Klein et al. [60] suggested that twinning occurred primarily during the nucleation stage. Berry and Skillman [4], on the other hand, believed that no twinning occurred during
nucleation but only during the subsequent crystal growth and Ostwald
ripening. Gutoff [7] found that in a continuous suspension crystallizer operated under steady state and high bromide ion excess conditions, twinned crystals did not appear until the crystal size reached
0.1 to 0.4 μm. One indication that twinning does not occur during
the first stages of precipitation is the observation by Hamilton and
Brady [62] that the two parallel twin planes in a tabular crystal
were separated by a distance of several hundred angstroms. In addi-

tion, multiply twinned AgBr crystals can be obtained from untwinned
crystals by Ostwald ripening [4].

Berry and Skillman [43] compared the growth rate of doubly twinned, singly twinned, and untwinned AgBr crystals and found that the
doubly twinned crystals grew much faster than the others. The higher
growth rate of the doubly twinned crystals was attributed to the presence of a reentrant angle (trough) where the twin plane intersects
the crystal surface [43,60,61]. Doubly twinned crystals always have
persistent rapid growth because there is no way the reentrant angles
can be eliminated by continued growth [63]. Singly twinned crystals
do not show rapid growth after the reentrant angles have grown them-
selves out of existence [4,43]. The rapid growth at the reentrant
angles has been attributed to the greater number of neighbor bonds in
a reentrant angle than on a flat surface [4,43,60,61,64].

The crystal number concentration (number/volume) has been shown
to influence the growth of AgX crystals. Klein and Moisar [31] reported that an excessive reduction in the concentration of crystals
could lead to the formation of new nuclei during the growth stage of
a double-jet precipitation. Similar results were also obtained by
Wey and Strong [28], who found a decrease in the critical growth rate
(at the supersaturation at which homogeneous nucleation just began
to occur) as crystal concentration was reduced. This effect was
attributed to an increase in the path and time of diffusion of ions
from regions of high local supersaturation to the crystal surface,
thus increasing the possibility of the formation of new nuclei
[5,28].

IV. OSTWALD RIPENING

A. Basic Concept

Any system is unstable thermodynamically until its free energy has
reached a minimum. Therefore, a two-phase system consisting of a

polydisperse precipitate within a solution phase is inherently unstable, since the large interfacial area contributes to the free
energy. Thus the free energy may be reduced by decreasing this area.
Ostwald [65] provided the first description of this process, which
therefore was later called "Ostwald ripening." Its driving force is
the difference in solubility between the polydisperse particles of
the precipitate, as given by the well-known Gibbs-Thomson equation
[Eq. (23)]. This solubility difference establishes a concentration
gradient, which leads to the transport of solute material through
the solution phase from the smaller crystals to the larger ones.
Thus the larger crystals grow at the expense of the smaller ones.
The rate of this ripening process is determined by the size distribution of the precipitate, its growth and dissolution kinetics, and
the transport properties of the solution phase.

The basic concept of Ostwald ripening is essentially the same
as that of the "Gibbs-Thomson effect" mentioned earlier for crystal
growth. However, Ostwald ripening normally refers to the case which
involves both the growth of the larger crystals and the dissolution
of the smaller crystals in a polydisperse population. Therefore,
Ostwald ripening may be regarded as a special case of the Gibbs-
Thomson effect.

Experiments designed to investigate Ostwald ripening are usually
done with a suspension of crystals under quasi-equilibrium conditions
(i.e., without the addition of reactant solutions). In this case,
the critical size, L^* $(= \Gamma_D/\ln S)$, is between that of the smallest
and the largest crystals. Those crystals with a size smaller than
L^* dissolve spontaneously because the effective driving force for
their growth is negative. The dissolved material forms a supersaturated solution with respect to the larger crystals and so deposits
on them. In some cases where reactant solutions are continuously
introduced, Ostwald ripening may still occur in the presence of complexing agents. For AgX crystals, typical complexing agents include
excess halide ions, NH_4OH, etc.

B. Kinetics

Empirical relationships describing the kinetics of Ostwald ripening have been given by Lyalikov [66] and Evva [67]. An equation which relates the number of crystals N and the ripening time t was suggested by Lyalikov [66]:

$$\frac{1}{N} = \frac{1}{N_0} + Kt \tag{28}$$

where N_0 is the number at the start of ripening and K is a rate constant which is nearly proportional to solubility. Evva [67] gave the dependence of average crystal volume $\overline{V}$ on the ripening time as

$$\overline{V} = \overline{V}_\infty [1 - \exp(-K_1 t)] \tag{29}$$

where $\overline{V}_\infty$ is the average crystal volume at the end of ripening and K_1 is the rate constant.

Detailed treatments of Ostwald ripening have been published [68–70]. Basically, Ostwald ripening is regarded as a two-step process involving successively the dissolution of material from the small crystals and the deposition of material onto the larger crystals. The intermediate step, which involves the transport of material through the "bulk solution" (i.e., the bulk region outside the diffusion layer of the crystals), is considered unimportant since Ostwald ripening is normally carried out in a well-agitated environment.

Wagner [69] assumed that growth (deposition) is a reciprocal of dissolution, i.e., both processes have the same kinetic expression. Taking either molecular diffusion [Eq. (14)] or surface integration [Eq. (11) with m = 1] as the rate-determining step for growth and dissolution, he derived the following equations to predict the average crystal size $\overline{L}$ under quasi-equilibrium conditions of Ostwald ripening:

$$\overline{L} \propto \left(\frac{TDC_s v^2 t}{R_g T} \right)^{1/3} \tag{30}$$

for diffusion control, and

$$\overline{L} \propto \left(\frac{TK_i C_s v^2 t}{R_g T} \right)^{1/2} \tag{31}$$

for surface integration control. Equations (30) and (31) predict that the average crystal volume $\overline{V}$ is proportional to the ripening time t for a diffusion-controlled mechanism and to $t^{3/2}$ for a surface integration-controlled mechanism. Lyalikov [71] and Berry and Skillman [35] observed a linear relationship between $\overline{V}$ and t for Ostwald ripening of AgBr crystals. These experimental results suggest that diffusion is the controlling mechanism.

Bogg et al. [72] used a mixture of two monodisperse AgBr crystals of widely different sizes to study Ostwald ripening under different solution conditions and crystal equilibrium habits. They also derived an equation for the "ripening time" (the time taken for the small crystals to dissolve completely). The rate of dissolution of the small crystals L_a was expressed as

$$\frac{dL_a}{dt} = K_a L_a^{\alpha}[C - C_e(L_a)] \tag{32}$$

and the rate of growth of the large crystals L_b was given by

$$\frac{dL_b}{dt} = K_b L_b^{\beta}[C - C_e(L_b)] \tag{33}$$

where K_a and K_b are rate constants, and α and β are constants whose values depend on the particular ripening mechanism. For example, $\alpha = \beta = 0$ represents surface integration control [Eq. (11) with m = 1] and $\alpha = \beta = -1$ represents diffusion control [Eq. (14)]. Here $\alpha = \beta$ if the mechanism of growth is the same as that of dissolution and $\alpha \neq \beta$ if the two processes have different mechanisms. Equations (32) and (33) can be combined with a mass balance equation and the

Gibbs-Thomson equation [Eq. (23)] to give an expression for the ripening time, t_R [72]:

$$t_R = \frac{L_a^{2-\alpha}}{(2-\alpha)C_s\Gamma_D K_a}\left[1 + \frac{2-\alpha}{4}\frac{K_a}{K_b}\frac{L_a^{\alpha-1}}{L_b^{\beta-1}}\frac{P_a}{P_b}\right] \tag{34}$$

where P_a/P_b is the ratio of weights of small to large crystals in the initial mixture. Equation (34) predicts that ripening time is inversely proportional to solubility C_s--a conclusion which is qualitatively supported by much experimental evidence [12,35,73]. The relative contributions to ripening time from growth and dissolution depend on the ratio K_a/K_b. The mechanisms of growth and dissolution can be inferred from the α and β values determined experimentally.

Bogg et al. [72] carried out ripening experiments in which small monodisperse crystals of AgBr were ripened onto larger ones in the presence of gelatin. They found that the ripening time was influenced both by the initial habit of the larger crystals and by the equilibrium crystal habit at the ripening conditions. When cubic crystals are present under conditions of octahedral stability and vice versa, deposition of material is energetically favored [53]. In this case ripening rates are determined by the rate of dissolution of the smaller crystals, provided that the amount of material deposited onto the larger crystals is not enough to cause significant changes in morphology. Changes in ripening time under these conditions reflect changes in solubility. On the other hand, if a crystal of given habit is present under conditions that favor that habit, then deposition of material is less energetically favored [53] and the ripening rates are determined by the rate of growth of the larger crystals. Consequently, ripening times are longer than would be predicted on a solubility basis [72]. It was also found that ripening time was independent of the habit of the dissolving crystals under all ripening conditions [72]. The above results are most compatible with a ripening mechanism based on diffusion-controlled dissolution and surface integration controlled growth ($\alpha = -1$, $\beta = 0$) [72]. That is, the rate-determining

step in dissolution is the diffusion of material away from the imme-
diate vicinity of the small crystals, and the growth of the larger
crystals is limited by the surface integration process [72].

A surface integration process appears to be the rate-determining
step for the growth of the larger AgBr crystals during Ostwald ripening. However, as indicated earlier, the growth of AgBr crystals during a double-jet precipitation is influenced both by diffusion and by
surface integration, with diffusion being more important [28]. These
inconsistent results may be explained by considering the great difference in supersaturation levels under the two "growth" conditions.
The supersaturation for crystal growth during precipitation is much
higher than that for the growth of larger crystals during Ostwald
ripening. For the former, the high supersaturation is due to the
continuous addition of reactant materials. For the latter, the low
supersaturation is a result of the slow dissolution of the smaller
crystals. It is generally believed that the effect of diffusion on
crystal growth is more pronounced under high supersaturation conditions (i.e., during precipitation) [37].

C. Activation Energy

Several attempts have been made to determine the activation energies
for the growth and dissolution of AgX crystals. Davies and Nancollas
[74] studied the rate of growth and the rate of dissolution of AgCl
seed crystals in the absence of gelatin. The solution supersaturation
or undersaturation was estimated by a conductivity technique. From
the temperature dependence of the rate constants, they concluded that
the activation energy for dissolution is 15.4 kcal mol^{-1} and for
growth it is zero. These values of activation energy suggest that
neither process is diffution-controlled. (For many electrolytes in
aqueous solution, the activation energy for diffusion is about 3.5
to 5 kcal mol^{-1} [75,76]). Similar measurements of the kinetics of
dissolution of AgCl crystals in the absence of gelatin were made by
Howard et al. [77]. They obtained an activation energy of 5 kcal
mol^{-1}, thus indicating that the dissolution of AgCl is

diffusion-controlled. Meehan and Beattie [12] studied the ripening of very dilute AgBr sols in the absence of gelatin using a light-scattering (turbidity) technique. They calculated an activation energy of 24 kcal mol^{-1} from a plot of $\log(dA/dt)$ versus $1/T$ (A = absorbance $\propto \bar{V}$).

Measurements of Ostwald ripening of AgX crystals in the presence of gelatin were obtained turbidimetrically by Berry and Skillman [35, 78] with a stopped-flow technique. In this scheme two solutions, one containing a suspension of small AgBr crystals and the other containing a complexing agent, were run together in an in-line mixer. The turbidity of the resulting suspension was measured at a small distance downstream from the mixer by a spectrophotometer set at a fixed wavelength. When the flow was stopped, the suspension in the spectrophotometer cell became more turbid with time due to an increase in $\bar{V}$ caused by Ostwald ripening. The activation energies were determined from a plot of $\log(d\bar{V}/dt)$ versus $1/T$. These investigators [35] found that activation energies for the ripening of fresh crystals in the presence of various complexing agents were within 2 kcal mol^{-1} of an average value of 15.7 kcal mol^{-1}. It was suggested that the ripening of freshly prepared crystals was dominated by structural imperfections which were present initially in the crystals, but which disappeared within a few minutes [34]. In a second study [78], the crystals were preaged for 10 min to eliminate the effect of structural imperfections. An activation energy of 25.4 kcal mol^{-1} was found for the ripening of AgBr cubic crystals in 0.1-N NH_4OH solution. For AgBr crystals with octahedral surfaces, the activation energy for ripening in 0.5-N KBr solution had a small value of 8.1 kcal mol^{-1}. The activation energy for AgCl crystals ripening in 0.25-N KCl was measured as 19.8 kcal mol^{-1} [78]. Berry and Skillman [35] concluded that the rates of Ostwald ripening were controlled by dissolution since their experimental values for the activation energy were in reasonable agreement with those for rates of solution of AgX crystals in coated photographic emulsions reported by James and Vanselow [79].

Claes and Borginon [27] studied Ostwald ripening of 0.1-μm AgBr crystals onto 0.7-μm AgBr cubes and octahedra in the presence of gelatin. The ripening pH was maintained at 9.5 by addition of NH_4OH, and the ratio of silver ion to bromide ion concentrations in solution at different temperatures was adjusted to 3.2×10^{-5}. The rate of Ostwald ripening was represented by the growth rate of the large crystals determined from electron micrographs. They found that at lower temperatures the octahedral crystals ripened faster than the cubic ones. However, at higher temperatures the opposite was true. The activation energies for Ostwald ripening were determined by plotting the logarithm of the growth rate of the large crystals versus 1/T, and values of 8 and 66 kcal mol^{-1} were obtained for octahedra and cubes, respectively.

The published activation energies for Ostwald ripening of AgX crystals vary widely. One possible explanation for this inconsistency is given here. Strictly speaking, the rate expression for a process (e.g., Ostwald ripening) should be divided into a temperature-dependent term and a composition-dependent term [80]:

$$\text{Rate} = f_1(\text{temperature}) \times f_2(\text{composition})$$

$$= k \times f_2(\text{composition}) \tag{35}$$

For many processes the temperature-dependent term (the rate constant k) is well represented by the Arrhenius equation:

$$k = k_0 \exp \frac{-\Delta E}{R_g T} \tag{36}$$

where k_0 is called the frequency factor, and ΔE is called the activation energy of the process, which can be determined from a plot of log k versus 1/T. The composition-dependent term is normally related to the concentration "driving force" for the process. For Ostwald ripening, the composition-dependent term may represent the supersaturation for the growth of large crystals or the undersaturation for the dissolution of small crystals. It is obvious that the composition-dependent term is influenced by the ripening conditions (temperature, excess halide ion concentration, halide type, presence of gelatin or complexing agent, etc.). Therefore, in order to obtain the "true"

activation energy for Ostwald ripening, one must separate the effect
of this composition-dependent term (i.e., the concentration driving
force for Ostwald ripening). However, most of the reported activation
energies [12,27,35,78,79] were determined from plots of the logarithm
of ripening rate versus 1/T, which is complicated by the effect of
the composition-dependent term. Furthermore, the activation energy
for Ostwald ripening is made up of contributions from dissolution and
growth, and the rate constants (k) for the two processes should be
measured separately at each temperature so that true activation energies can be calculated.

Based on the above considerations, it is not surprising that the
reported activation energies for Ostwald ripening are so wide ranging
and inconsistent, since they do not represent the "true" values. Attempts to separate the effect of the composition-dependent term are
hindered by the difficulties in accurately measuring the bulk solute
concentration during Ostwald ripening of AgX crystals. Therefore,
the previous approaches [12,27,35,78], which made use of the activation energy measurements to elucidate the mechanism of Ostwald ripening, appear to have little fundamental basis.

D. Factors Affecting Ostwald Ripening

Ostwald ripening of AgX crystals occurs more rapidly at high-solubility
conditions (higher temperature, excess halide ion, presence of complexing agent, etc.) and when there is a wide distribution of crystal
sizes. For example, Ostwald ripening is very important in single-
jet precipitations due to the presence of a large excess of halide
ions. Because of Ostwald ripening during single-jet precipitation,
the total number of crystals in suspension is believed to decrease
and the width of the distribution to increase as a function of time
[52].

Berry [51] studied Ostwald ripening of AgBr crystals at 50°C
and pAg 9.9 under quasi-equilibrium conditions. Before ripening the
crystals had flat faces, but after 200 min of ripening they were
nearly spherical. The spherical appearance of slowly ripening crystals

is entirely logical. Ostwald ripening requires parts of some or of all the crystals to dissolve. Dissolution most likely will occur at the sharp corners and edges since fewer neighbor bonds have to be broken. This etching process can continue and result in rounded crystals.

NH_4OH has some interesting effects on Ostwald ripening of AgX crystals. Generally, NH_4OH enhances the ripening rate [53,72] due to an increase in total solubility of AgX via the formation of complex ions [81]. However, at high NH_4OH concentrations the ripening rate is extremely slow whatever the initial habit of the crystals [72,82]. This retardation of Ostwald ripening was attributed to the stabilization of the crystal faces by a large amount of adsorbed $Ag(NH_3)_2^+$ [83].

The effect of pH on Ostwald ripening was examined by several investigators [73,84]. Klein et al. [73] observed that the rate of ripening of AgBr in an inert bone gelatin solution increased monotonically with pH at pAg $\simeq$ 9 (i.e., pBr $\simeq$ 2, excess bromide ion) and decreased with increasing pH at pAg $\simeq$ 3 (excess silver ion). The differences in ripening rate were ascribed to electrostatic interactions between the protective colloid and the AgX surface [73]. When the crystal is positively charged (at pAg 3), gelatin is repulsed by the crystal at low pH (since gelatin is cationic at this condition). This causes a decrease of the overall binding forces. The crystal surface is less covered so that Ostwald ripening is facilitated. At high pH values, gelatin is in an anionic state and is attracted by the positively charged crystal. The surface is blocked so that Ostwald ripening is retarded. The same considerations can be applied to high pAg where the crystal bears a net negative charge due to halide ion adsorption. This leads to the opposite behavior in which the ripening rate is directly related to pH.

Cohen et al. [84] turbidimetrically determined the dependence of the ripening rate of AgBr crystals on pH at different levels of bromide ion excess. They concluded that although ripening rates of AgBr were pH-dependent, the charge of the gelatin was probably not the determining factor. At low bromide ion concentrations the pH dependence was attributed to the amine moieties of the gelatin, whereas at higher bromide ion excess the rate enhancement at low pH appeared to be independent of the gelatin [84].

Oppenheimer et al. [85] studied the effect of some cationic surfactants on the Ostwald ripening of AgBr. The rate of ripening was found to be directly related to the adsorptivity of the surfactants at the crystal surface. Enhancement of ripening by adsorbed cationic surfactants was explained by bromide counterion adsorption and subsequent formation of mobile $AgBr_2^-$ in the electrical double layer [85].

NOMENCLATURE

B = Constant given in Eqs. (7)-(9)

C = Solute concentration in the bulk solution, mol/vol

C_e = Solute concentration in equilibrium with a solid phase, mol/vol

C_i = Solute concentration at the solid-solution interface, mol/vol

C_s = Solubility (solute concentration in equilibrium with a flat solid phase), mol/vol

D = Diffusivity, area/time

ΔE = Activation energy, energy/mol

G = Rate of increase of crystal size, length/time

g = Size-independent growth rate defined by Eq. (26), length/time

K = Rate constant given in Eq. (28), 1/time

K_a = Rate constant given in Eq. (32)

K_b = Rate constant given in Eq. (33)

K_1 = Rate constant given in Eq. (29), 1/time

K_i = Surface integration constant given in Eq. (11)

K_{sp} = Solubility product, $(mol/vol)^2$

k = Rate constant given in Eq. (35)

k_d = Mass transfer coefficient, length/time

k_0 = Frequency factor given in the Arrhenius equation, Eq. (36)

L = Size (edge length or diameter) of crystal, length

L_a = Size of small crystal population, length

L_b = Size of large crystal population, length

$\overline{L}$ = Average size of crystals, length

L^* = Critical size of crystal or nucleus defined by Eq. (27), length

m = "Order" of surface integration given in Eq. (11)

N = Number of crystals during Ostwald ripening, number

N_0 = Number of crystals at the start of Ostwald ripening, number

P_a = Total mass of small crystal population, mol

P_b = Total mass of large crystal population, mol

R = Rate of reactant addition, mol/(time)(vol)

Re = Reynolds number ($u_s L \rho / \mu$), dimensionless

R_g = Gas constant, energy/(mol)(temp)

S = Supersaturation ratio (C/C_s), dimensionless

Sc = Schmidt number ($\mu / \rho D$), dimensionless

Sh = Sherwood number ($k_d L/D$), dimensionless

T = Absolute temperature

t = Time

t_R = Ripening time defined by Eq. (34), time

u_s = Particle slip velocity, length/time

u_t = Particle terminal velocity, length/time

$\overline{V}$ = Average crystal volume during Ostwald ripening, vol

$\overline{V}_\infty$ = Average crystal volume at the end of Ostwald ripening, volume

v = Molar volume, vol/mol

Z = Total number of nuclei per unit volume defined by Eq. (6), number/vol

Z_0 = Total number of initial nuclei per unit volume, number/vol

α = Constant given in Eq. (32)

β = Constant given in Eq. (33)

β_n = Stability constants, n = 1, 2, 3, ...

Υ = Surface energy, energy/area

Γ_D = Capillary constant defined by Eq. (24), length

δ = Thickness of the diffusion layer around a crystal surface, length

ε = Growth rate parameter defined by Eq. (22), 1/length

μ = Viscosity, mass/(time)(length)

ν = Stoichiometric factor given in Eq. (24)

ρ = Solution density, mol/vol

ρ_c = Crystal density, mol/vol

σ = Supersaturation parameter $[(C_i - C_e)/C_e]$, dimensionless

σ_c = Constant given in Eq. (10)

ACKNOWLEDGMENTS

The author is indebted to Dr. J. P. Terwilliger for many helpful discussions and suggestions during the preparation of the manuscript. Acknowledgment is given to Drs. R. E. Bacon, J. I. Cohen, M. J. Hazard, and A. H. Herz for useful comments. The author would also like to thank Dr. W. R. Wilcox (editor of this volume and series) for critically reviewing the manuscript. Finally, the author wishes to express his gratitude to the management of the Kodak Research Laboratories for their encouragement and support in writing this review.

REFERENCES

1. C. R. Berry, in *The Theory of the Photographic Process*, 4th ed. (T. H. James, ed.), Macmillan, New York, 1977, Chap. 1, 3.

2. G. F. Duffin, *Photographic Emulsion Chemistry*, Focal, London, 1966.

3. V. L. Zelikman and S. M. Levi, *Making and Coating Photographic Emulsions*, Focal, London, 1964.

4. C. R. Berry and D. C. Skillman, *J. Photogr. Sci.*, *16*, 137 (1968).

5. E. Moisar, in *VIth Int. Conf. Corpuscular Photography*, Florence, July 19–23, 1966 (M. Della Corte, ed.), Edizioni C. E. P. I., Roma.

6. Y. A. Breslav and V. A. Uksusova, *Zh. Nauchn. Prikl. Fotogr. Kinematogr.*, *19*, 296 (1974).

6a. F. Claes and R. Berendsen, *Photogr. Korresp.*, *101*, 37 (1965).

7. E. B. Gutoff, *Photogr. Sci. Eng.*, *14*, 248 (1970); *15*, 189 (1971).

8. J. S. Wey, J. P. Terwilliger, and A. D. Gingello, *AIChE Symp. Ser.*, *76(193)*, 34 (1980).

9. A. D. Randolph and M. A. Larson, *Theory of Particulate Processes*, Academic, New York, 1971.

10. C. R. Berry and D. C. Skillman, *J. Phys. Chem.*, *68*, 1138 (1964).

11. R. W. Berriman, *J. Photogr. Sci.*, *12*, 121 (1964).

12. E. J. Meehan and W. J. Beattie, *J. Phys. Chem.*, *65*, 1522 (1961).

13. V. M. Nefedchenkov and M. I. Rudenko, *Prikl. Fiz. Tverd. Tela.*, 209 (1973).

14. M. Volmer, *Kinetik der Phasenbildung*, Edwards Bros., Ann Arbor, Mich., 1945.

15. J. A. Christiansen and A. E. Nielsen, *Acta Chem. Scand.*, 5, 673 (1951).

16. D. Turnbull and J. C. Fisher, *J. Chem. Phys.*, *17*, 71 (1965).

17. A. E. Nielsen, *Acta Chem. Scand.*, *11*, 1512 (1957).

18. A. C. Zettlemoyer, *Nucleation*, Marcel Dekker, New York, 1969.

19. A. E. Nielsen, *Kinetics of Precipitation*, Pergamon, Oxford, 1964.

20. A. G. Walton, *The Formation and Properties of Precipitates*, Interscience, New York, 1967.

21. J. Estrin, in *Preparation and Properties of Solid State Materials*, Vol. 2 (W. R. Wilcox, ed.), Marcel Dekker, New York, 1976, chap. 1.

21a. G. D. Botsaris, in *Industrial Crystallization*, (J. W. Mullin, ed.), Plenum, New York, 1976, p. 3.

22. C. W. Davies and A. L. Jones, *Discuss. Faraday Soc.*, 5, 103 (1949).

23. D. H. Klein, L. Gordon, and T. H. Walnut, *Talanta*, *3*, 177 (1959).

24. J. W. Mullin, *Crystallization*, 2nd ed., Butterworths, London 1972.

25. C. R. Berry, *Photogr. Sci. Eng.*, *20*, 1 (1976).

26. E. B. Gutoff, F. R. Cottrell, G. Margolis, E. G. Denk, and C. H. Wallace, *Photogr. Sci. Eng.*, *18*, 8 (1974).

27. F. H. Claes and H. Borginon, *J. Photogr. Sci.*, *21*, 155 (1973).

28. J. S. Wey and R. W. Strong, *Photogr. Sci. Eng.*, *21*, 14 (1977).

29. P. F. Ipatov, *Usp. Nauchn. Fotogr.*, *1*, 39 (1955).

30. C. Wagner, private communication cited in E. Klein and E. Moisar, Mitt. Forschungslabor, Agfa-Gevaert, Bd. IV, 38 (1964).

31. E. Klein and E. Moisar, *Ber. Bunsenges. Phys. Chem.*, *67*, 349 (1963).

32. V. G. Loginov and N. B. Denisova, *Zh. Nauchn. Prikl. Fotogr. Kinematogr.*, *20*, 231 (1975).

33. T. Sakaguchi and T. Miura, *J. Soc. Photogr. Sci. Tech. Jap.*, *36*, 160 (1973).

34. C. R. Berry and D. C. Skillman, *J. Cryst. Growth*, 2, 141 (1968).

35. C. R. Berry and D. C. Skillman, *J. Phys. Chem.*, *70*, 1871 (1966).

36. M. Ohara and R. C. Reid, *Modeling Crystal Growth Rates from Solution*, Prentice-Hall, Englewood Cliffs, N.J., 1973.

37. J. Garside, in *1976 Crystal Growth and Materials* (E. Kaldis and H. J. Scheel, eds.), North-Holland, Amsterdam, 1977, Chap. II.2. p. 2.

38. F. C. Frank, *Discuss. Faraday Soc.*, 5, 48 (1949).

39. W. K. Burton, N. Cabrera, and F. C. Frank, *Philos. Trans. R. Soc. London*, 243, 299 (1951).

39a. W. R. Wilcox, *J. Cryst. Growth*, 37, 229 (1977).

40. R. B. Bird, W. E. Stewart, and E. N. Lightfoot, *Transport Phenomena*, Wiley, New York, 1960.

41. D. N. Miller, *Ind. Eng. Chem. Process Des. Dev.*, 10, 365 (1971).

42. N. Frössling, *Gerlands Beitr. Geophys.*, 52, 170 (1938).

43. C. R. Berry and D. C. Skillman, *Photogr. Sci. Eng.*, 6, 159 (1962).

44. C. R. Berry and D. C. Skillman, *J. Phys. Chem.*, 67, 1827 (1963).

45. A. Hirata and S. Hohnishi, *Bull. Soc. Photogr. Sci. Jap.*, 16, 1 (1966).

46. J. S. Wey and R. W. Strong, *Photogr. Sci. Eng.*, 21, 248 (1977).

47. R. W. Strong and J. S. Wey, *Photogr. Sci. Eng.*, 23, 344 (1979).

48. B. V. Enüstün and J. Turkevich, *J. Amer. Chem. Soc.*, 82, 4502 (1960).

48a. T. Sugimoto and G. Yamaguchi, *J. Colloid Interface Sci.*, 57, 400 (1976).

49. J. S. Wey and J. Estrin, *AIChE Symp. Ser.*, 68(121), 74 (1972).

50. J. S. Wey and J. P. Terwilliger, *AIChE J.*, 20, 1219 (1974).

51. C. R. Berry, *Photogr. Sci. Eng.*, 18, 4 (1974).

52. G. Margolis and E. B. Gutoff, *AIChE J.*, 20, 467 (1974).

53. E. Moisar and E. Klein, *Ber. Bunsenges. Phys. Chem.*, 67, 949 (1963).

54. I. Langmuir, *J. Amer. Chem. Soc.*, 38, 2221 (1916).

54a. W. L. Gardner, D. P. Wrathall, and A. H. Herz, *Photogr. Sci. Eng.*, 21, 325 (1977).

54b. F. H. Claes, J. Libeer, and W. Vanassche, *J. Photogr. Sci.*, 21, 39 (1973).

55. C. R. Berry, S. J. Marino, and C. F. Oster, *Photogr. Sci. Eng.*, 5, 332 (1961).

56. W. Markocki and A. Zaleski, *Photogr. Sci. Eng.*, 17, 289 (1973).

57. F. H. Claes and W. Peelaers, *Photogr. Korresp.*, 103, 161 (1967).

58.   A. Hirata and S. Hohnishi, *J. Photogr. Sci. Jap.*, *36*, 359 (1973).

59.   F. H. Claes and W. Peelaers, *Photogr. Sci. Eng.*, *12*, 207 (1968).

60.   E. Klein, H. J. Metz, and E. Moisar, *Photogr. Korresp.*, *99*, 99 (1963).

61.   R. W. Berriman and R. H. Herz, *Nature*, *180*, 293 (1957).

62.   J. F. Hamilton and L. E. Brady, *J. Appl. Phys.*, *35*, 414 (1964).

63.   D. R. Hamilton and R. G. Seidensticker, *J. Appl. Phys.*, *31*, 1165 (1961).

64.   T. Sugimoto and G. Yamaguchi, *J. Cryst. Growth*, *34*, 253 (1976).

65.   W. Ostwald, *Analytische Chemie*, 3rd ed., Engelmann, Leipzig, 1901.

66.   K. S. Lyalikov, in *Statistical Phenomena on Heterogeneous Systems*, Academy of Science of USSR, 1949, p. 174.

67.   F. Evva, *Z. Wiss. Photogr. Photophys. Photochem.*, *47*, 39, 84 (1952).

68.   I. M. Lifshitz and V. V. Slyozov, *J. Phys. Chem. Solids*, *19*, 35 (1961).

69.   C. Wagner, *Z. Elektrochem.*, *65*, 581 (1961).

70.   M. Kahlweit, *Advan. Colloid Interface Sci.*, *5*, 1 (1975).

71.   K. S. Lyalikov, *Usp. Nauchn. Fotogr.*, *5*, 39 (1957).

72.   T. G. Bogg, M. J. Harding, and D. N. Skinner, *J. Photogr. Sci.*, *24*, 81 (1976).

73.   E. Klein, E. Moisar, and E. Roche, *J. Photogr. Sci.*, *19*, 55 (1971).

74.   C. W. Davies and G. H. Nancollas, *Trans. Faraday Soc.*, *51*, 818 (1955).

75.   C. E. Davis and E. T. Oakes, *J. Amer. Chem. Soc.*, *44*, 464 (1922).

76.   L. L. Bircumshaw and A. C. Riddiford, *Q. Rev.*, *6*, 157 (1952).

77.   T. R. Howard, G. H. Nancollas, and N. Purdie, *Trans. Faraday Soc.*, *56*, 278 (1960).

78.   C. R. Berry and D. C. Skillman, *Photogr. Sci. Eng.*, *13*, 69 (1969).

79.   T. H. James and W. Vanselow, *Photogr. Sci. Tech. II*, *2*, 135 (1955).

80.   O. Levenspiel, *Chemical Reaction Engineering*, Wiley, New York, 1962, p. 22.

81.   H. Z. Ammann-Brass, *Z. Wiss. Photogr. Photophys. Photoch. .*, *55*, 103 (1961).

82.   G. Danguy, *Bull. Soc. Roy. Sci. Liege*, *32*, 87, 790 (1963).

83.   E. Klein and E. Moisar, *J. Photogr. Sci.*, *12*, 242 (1964).

84. J. I. Cohen, W. L. Gardner, and A. H. Herz, *Advances in Chemistry Series*, no. 145, 198 (1975).

85. L. E. Oppenheimer, T. H. James, and A. H. Herz, in *Particle Growth in Suspensions* (A. L. Smith, ed.), Academic, London, 1973, p. 159.

Chapter 3

HEAT TRANSFER IN CZOCHRALSKI CRYSTAL GROWTH

Nobuyuki Kobayashi

Department of Applied Physics
Faculty of Engineering
Toyama University
Takaoka, Toyama, Japan

I. INTRODUCTION

The Czochralski technique is one of the most important methods for the
growth of single crystals. The growth process is basically simple.
The raw material is melted in an appropriate crucible. The single-
crystal seed is dipped vertically into the melt and wetted completely.
After the seed and melt reach steady state, the seed is slowly with-
drawn upward. Then one is able to obtain a single crystal under the
seed. A crystal is easily grown if the melt is stable and can be contained in a nonreactive crucible. Thus the technique is widely employed for various materials, including metals, semiconductors, and
insulators and ranging from low-melting-point materials to high-melting-
point ones. Crystals of high quality are often obtainable by improvement of the furnace.

Czochralski growth, and other methods of crystal growth from the melt, are mainly controlled by heat transfer. Heat applied to the crucible is almost totally consumed for the maintenance of the solid-liquid interface at the melting point and the melt at a temperature above the melting point. In Czochralski growth, the heat is transferred to the crucible by a heat source and dissipated to the heat sinks around the crucible, keeping the solid-liquid interface roughly at the level of the bulk melt free surface. When the crystal is growing, the latent heat of solidification, which is proportional to the growth rate of the crystal, is delivered at the solid-liquid interface. The crystal growth is controlled by the delivery of the sinks through the crystal and/or the melt. If the latent heat is easily delivered to heat sinks through the crystal, high growth rate is attainable.

The heat applied to the crucible by a heat source and the latent heat delivered at the interface by the solidification are superposed. By considering conservation of heat at the interface, it is seen that the latent heat rate is given by the difference between the heat conducted away from the interface to the crystal and that transferred from the melt to the interface. The heat flow near the interface is not easily calculated because it depends on many internal and external factors. Thus control of the growth is based mainly on experience. For example, if one somehow increases the heat flow from the interface to the crystal or decreases the heat flow from the melt to the interface, one is able to increase the diameter of the crystal. Usually, the diameter is controlled by varying the melt temperature via the applied power to the crucible, which varies the heat flow from the melt to the interface. The pulling rate may also be varied.

Crystals of high quality are indispensable for research in solid state physics and for application to electronics devices. Larger, more uniform, and better crystals are needed and many attempts have been made to improve crystal quality. For example, dislocation-free crystals are obtained by necking in during the initial stage of growth. Residual strain-free crystals are also obtained by controlling the interface shape; a planar interface shape is generally favorable

during crystal growth. Cracks which represent catastrophic damage
to the crystal are avoidable by reducing the temperature gradient in
the crystal. Presently, their relation to the heat transfer is not
well known because of the difficulty in the estimation of the thermal
stress generated in the crystal growth. The inhomogeneity of the
solute segregation is determined by mass transfer in the melt; crystal
rotation has proved to be quite useful to obtain radially uniform
crystals.

As mentioned above, crystal growth depends heavily not only on
heat transfer but also on mass transfer. In order to obtain crystals
of high quality it is necessary to control both. In the melt, they
are influenced by fluid flow (convection). Fortunately the analysis
of heat and mass transfer is a sophisticated and well-developed field
quite useful to the understanding of the crystal growth. The results
developed for heat transfer during Czochralski growth mainly are reviewed here.

II. SCHEMATIC DESCRIPTION OF HEAT FLOW DURING CZOCHRALSKI CRYSTAL GROWTH

Heat flow in the furnace is very complex, as shown schematically in
Fig. 1. The heat supplied to the crucible is transferred to the heat
sinks around the furnace in various manners. The heat is partly conducted or radiated directly from the crucible to the melt and is partly radiated from the crucible to the crystal. The heat conducted to
the melt is transferred to the interface and melt surface by conduction and by convection. Heat is dissipated from the melt free surface
to the atmosphere by gaseous convection and by radiation. The heat
transferred to the crystal through the interface is conducted to the
crystal surface, where the heat is dissipated by gaseous convection
and by radiation. Sometimes the heat dissipation from the crystal
surface and melt free surface is controlled by a reflector or afterheater placed around the crystal. If the crystal is growing at high
temperature, heat transfer by radiation becomes dominant among the
three heat transfer modes, conduction, convection, and radiation. In
the growth of some oxide crystals, radiative heat transfer is important

in the crystal and in the melt. When the crystal is growing, latent
heat is delivered by solidification at the solid-liquid interface and
is conducted to the crystal, and to the melt if it is supercooled.
The control of the growth depends on whether the latent heat is easily
delivered to the heat sinks or not by a change in the thermal
conditions.

To accomplish quantitative control of Czochralski growth, the
thermal variables of the furnace are examined in detail. These are:

1. The factors which can generally be determined before the crystal
    growth are the geometric factors of the elements in the furnace
    such as the heat source, crucible, crucible support, afterheater,
    reflector, pulling rod, etc.
2. The factors which are changeable during crystal growth are mechani-
    cal factors such as the flow rate of the gas, the pulling rate,
    the crystal rotation rate, the crucible rotation rate, the input
    power, etc.
3. The factors which vary during the growth of a given crystal are
    the crystal size, the height of the melt, and meniscus shape.
    Strictly speaking, these factors are determined by the time-
    dependent heat transfer. These factors can be regarded as pre-
    determined variables under the present stage of our knowledge on
    the heat transfer.
4. The factors which are important before and after the crystal
    growth are seeding rate, cooling rate, etc. Fast seeding some-
    times results in generation of dislocations in the seed by ther-
    mal shock. Cooling rate is controlled for the crystal with any
    phase transition temperature above room temperature.
5. Among material properties, thermal properties are of prime impor-
    tance. These are thermal conductivity, specific heat, density,
    expansion coefficient, emissivity of radiation, absorption coeffi-
    cient of radiation, melting point, and latent heat. The growth
    kinetics, cleavage behavior, dislocation density of the seed,
    and any phase transition temperature can also be important factors.
    For the melt, diffusion coefficient, viscosity, surface tension,
    and contact angle are important.

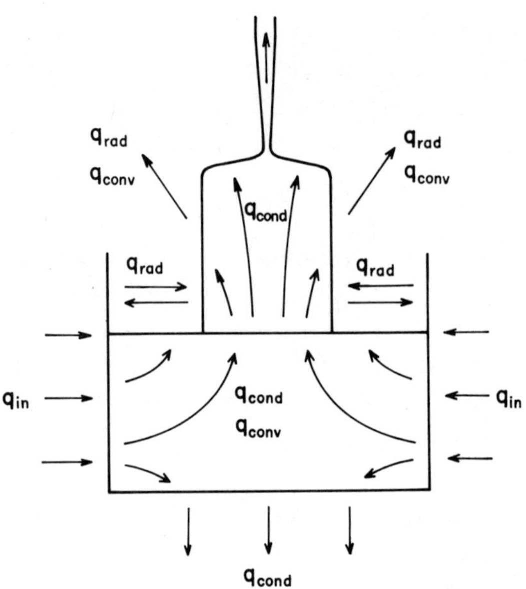

FIG. 1.   Schematic heat flow during Czochralski crystal growth.

III.   POWER REQUIREMENT

For stable crystal growth, one needs to obtain a melt in a crucible.
If the material has a low melting point, a melt is usually easily obtained.   However if the material has a high melting point, there may
be considerable difficulty in finding a relevant crucible which is
nonreactive to the melt and in finding a suitable heating device which
is powerful enough to melt the material.

For high-temperature crystal growth, the prime heat transfer
mechanism is radiation.   If the melt temperature is near the melting
point and a heat flux from the surroundings is neglected, the heat
flux into the surroundings is roughly given by

$$q_\ell = \varepsilon_\ell \sigma \, T_m^4 \tag{1}$$

where ε_ℓ is the emissivity of the melt, σ is the Stefan-Boltzmann
constant, and T_m is the melting point.   If a crystal is present, heat
is also conducted from the interface into the crystal and is dissipated

by radiation into the surroundings from crystal surface. The heat
flux from the interface into the crystal with constant diameter is
approximately [1]

$$q_s = \frac{k_s T_m}{a}\left(\frac{0.8\varepsilon_s \sigma T_m^3 a}{k_s}\right)^{1/2}$$

where k_s is the thermal conductivity of the solid, a is the crystal
radius, and ε_s is the emissivity of the solid. This expression will
be derived in Sec. V.

In Table 1, the heat flux q_ℓ and q_s are shown for various materials, including the metals, semiconductors, and oxides. In Table 2,
various devices or heat sources are shown with their heat transfer
intensities [1]. The comparison between heat flux and the heat transfer intensity is quite suggestive for the choice of a suitable device
or crystal growth method.

In Czochralski crystal growth, the maximum attainable melt temperature may be limited by the crucible, because it should not react with
the melt. For oxide melts, iridium with a melting point of 2443°C is
often used, with the highest melting temperature recorded being 2135°C
for $MgAl_2O_4$ [2]. Above a temperature of 2443°C, crucible-free techniques must be used [3]. In these techniques the material is only
partially melted to form a pool of melt on the top-central portion
of the material. If the melt pool is stable, crystal pulling is
possible

IV. BASIC EQUATIONS

If the crystal is neither growing nor melting, the interface shape
is given by the melting point isotherm. For crystal growth, supercooling is required at the interface to provide the driving force
for the growth. The interface may no longer be an isotherm, but a
nonisothermal surface reflecting the anisotropy of the crystal lattice and growth kinetics. The state near the interface depends on
the variables which determine the free energy of the phase, such as
temperature, solute concentration, the interface shape, a mechanical

TABLE 1

Heat Flux for Various Materials

	Cu	W	Ge	Si	SBN	Cr_2O_3
Melting point T_m (K)	1356	3653	1212	1685	1470	2538
Thermal conductivity k_s (W cm^{-1} K^{-1})	3.3	0.895	0.174	0.220	0.008	0.04
Emissivity ε_s	0.056	0.36	0.59	0.46	0.88	0.8
Crystal radius[a] a (cm)	1	1	1	1	1	1
Heat flux[b] q_ℓ (W cm^{-2})	1.07	363	7.17	21.0	23.3	188
q_s (W cm^{-2})	62.0	945	34.7	80.0	14.8	124
Total heat flux $Q_s = \pi a^2 q_s$ (W)	195	3060	109	248	46.5	388

[a]Assumed.

[b]$\varepsilon_\ell = \varepsilon_s$ is assumed.

Sources: Refs. 1, 71.

TABLE 2

Various Heat Sources

Source	Mechanism	Intensity (kW cm^{-2})
Air-fuel flames	Gaseous convection	0.04
Oxy-fuel flames	Gaseous convection	0.3
Constricted flames	Gaseous convection	3.2
Laminar argon plasma	Gaseous convection	0.2
Atomic hydrogen flame	Gaseous convection	1.3
Argon plasma jets	Gaseous convection	4.5
DC arcs	Electron transfer and gaseous convection	16
Electron beams[a]	Electron transfer	10^3
Image devices	Photon transfer	1.5
Lasers[a]	Photon transfer	10^3
Tungsten soldering iron	Phonon transfer	7

[a]At very high intensities catastrophic vaporization occurs, limiting further increase.
Source: Ref. 1; reprinted with permission.

balance at the interface, etc., and on the mechanism of crystal growth,
i.e., the attachment mechanisms of atoms on the interface. These
variables are now described.

A. Temperature T(x, y, z)

If crystal growth is mainly controlled by heat transfer, temperature
is one of the most important variables. The heat flux vector $\vec{q}$ =
(q_x, q_y, q_z), the rate at which heat is transferred across any surface per unit area per unit time, is generally expressed by the components of the temperature gradient [4]. For an anisotropic solid,

$$\begin{pmatrix} q_x \\ q_y \\ q_z \end{pmatrix} = - \begin{pmatrix} k_{11} & k_{12} & k_{12} \\ k_{21} & k_{22} & k_{23} \\ k_{31} & k_{32} & k_{33} \end{pmatrix} \begin{pmatrix} \partial T/\partial x \\ \partial T/\partial y \\ \partial T/\partial z \end{pmatrix} \tag{3}$$

where the quantities k_{ij} are the components of a second-order thermal
conductivity tensor. The heat transfer equation is given by

$$\rho c_p \frac{\partial T}{\partial t} = - \left(\frac{\partial q_x}{\partial x} + \frac{\partial q_y}{\partial y} + \frac{\partial q_z}{\partial z} \right) = - \nabla \cdot \vec{q} \tag{4}$$

where ρ is the density and c_p is the specific heat at constant pressure. Equation (3) is frequently simplified if the axes are chosen
in appropriate crystallographic directions because of crystalline symmetry. The simplified thermal conductivity tensor for crystal systems are given in Appendix A. For an isotropic solid, Eq. (4) becomes

$$\rho c_p \frac{\partial T}{\partial t} = \nabla (k \, \nabla T) \tag{5}$$

where k is the thermal conductivity of the medium. If the solid is
moving with a velocity $\vec{v}$ = (v_x, v_y, v_z), Eq. (2) is replaced by

$$\rho c_p \left(\frac{\partial T}{\partial t} + \vec{v} \cdot \nabla T \right) = \nabla (k \, \nabla T) \tag{6}$$

In a moving compressible fluid, heat transfer equation is given by [5]

$$\rho c_p \left(\frac{\partial T}{\partial t} + \vec{v} \cdot \nabla T \right) = \nabla (k \, \nabla T) + \left(\frac{\partial \ln V}{\partial \ln T} \right)_\rho \left(\frac{\partial p}{\partial t} + \vec{v} \cdot \nabla p \right) + \mu \phi \tag{7}$$

where μ is the viscosity, V is the volume, and p is the pressure. ϕ is the dissipation function defined by

$$\phi = 2 \left[\left(\frac{\partial v_x}{\partial x} \right)^2 + \left(\frac{\partial v_y}{\partial y} \right)^2 + \left(\frac{\partial v_z}{\partial z} \right)^2 \right] + \left(\frac{\partial v_y}{\partial x} + \frac{\partial v_x}{\partial y} \right)^2 + \left(\frac{\partial v_z}{\partial y} + \frac{\partial v_y}{\partial z} \right)^2$$

$$+ \left(\frac{\partial v_x}{\partial z} + \frac{\partial v_z}{\partial x} \right)^2 - \frac{2}{3} (\nabla \cdot \vec{v})^2 \tag{8}$$

Usually, the last term in Eq. (7) is safely neglected because it is small with respect to the others. For a fluid at constant pressure or fluids with a density independent of temperature, Eq. (7) becomes [5]

$$\rho c_p \left(\frac{\partial T}{\partial t} + \vec{v} \cdot \nabla T \right) = \nabla (k \, \nabla T) + \mu \phi \tag{9}$$

At the boundary surfaces, temperature may be specified by various conditions.

1. Prescribed surface temperature:
If the temperature at the crucible is controlled to be constant, it may may be specified by $T = T_c$.

2. Prescribed heat flux across the surface:
The heat flux at the surface may be given by $q = q_0$. This is equivalent to specifying the temperature gradient. For an anisotropic medium, the heat flux normal to the surface is given by [4]

$$\vec{n} \cdot \vec{q} = \vec{n} \cdot (k_n \nabla T) \tag{10}$$

where $\vec{n} = (n_1, n_2, n_3)$ is the unit vector normal to the surface and $k_n = k_{11} n_1^2 + k_{22} n_2^2 + k_{33} n_3^2 + 2 k_{12} n_1 n_2 + 2 k_{23} n_2 n_3 + 2 k_{13} n_1 n_3$. For an isotropic medium, $k_n = k$. If there is no heat flux across the surface, $q = 0$.

3. Heat flux transferred by fluid convection:
Sometimes the heat flux at a solid-gas or liquid-gas boundary is related to the temperature difference between the boundary and the bulk fluid [6]:

$$q = h(T - T_b)$$

where h is the heat transfer coefficient and T_b is the temperature in the bulk fluid. This relation is sometimes referred to as "Newton's law of cooling." If heat is transferred by free convection in a gaseous fluid, the heat transfer coefficient from the surface is usually given by [6]

$$\frac{hL_c}{k} = C_0 Ra^{1/n} \tag{12}$$

where L_c is a characteristic length, C_0 is a constant, and Ra is the Rayleigh number defined by using the properties of the gas:

$$Ra = \frac{g\beta \, \Delta T L_c^3}{\nu^2} \left(\frac{\nu}{\alpha_t}\right) \tag{13}$$

where ΔT is the temperature difference between the crystal and the atmosphere. If the convection is laminar, n = 4, and if turbulent, n = 3 [6].

4. Heat flux transferred by blackbody radiation:

A body at the absolute temperature T surrounded by the blackbody at temperature T_a will radiate heat at the rate [6]

$$q = \varepsilon\sigma(T^4 - T_a^4) \tag{14}$$

where ε is the emissivity of the surface and σ is the Stefan–Boltzmann constant. If the temperature difference $T - T_a$ is not large, this is approximately given by

$$q = 4\varepsilon\sigma T_a^3 (T - T_a) \tag{15}$$

If a body is surrounded by many gray surfaces with different temperatures T_i (i = 2 to N), the heat radiated from its surface must be obtained by gray enclosure theory [6]. This is roughly given by consideration of direct radiative interchange

$$q = \sum_{i=2}^{N} \frac{\varepsilon_1 \varepsilon_i f_{1i} (T_1^4 - T_i^4)}{1 - (1 - \varepsilon_1)(1 - \varepsilon_i) f_{1i} f_{i1}} \tag{16}$$

where ε_i is the emissivity of the ith surface and f_{ij} is the geometric view factor from the ith surface to the jth surface as defined in Appendix B.

5. Continuities of temperature and heat flux at the boundary between two media:

At the solid-liquid interface,

$$T_s = T_\ell \quad \text{and} \quad q_{n,s} = q_{n,\ell} \tag{17}$$

If the heat is generated by solidification at the interface,

$$T_s = T_\ell \quad \text{and} \quad q_{n,s} = q_{n,\ell} + \rho_s \Delta Hf \tag{18}$$

where ρ_s is the density of solid, ΔH is the latent heat, and f is the growth rate.

B. Solute Concentration C(x, y, z)

The solute concentration in a binary mixture satisfies a diffusion equation similar to that governing the temperature for a fluid of constant mass density [7]:

$$\frac{\partial C}{\partial t} + \vec{v} \cdot \nabla C = \nabla (D \nabla C) \tag{19}$$

where D is the diffusion coefficient. This equation is usually used for diffusion in dilute liquid solutions.

The boundary conditions used for diffusion equation are similar to those used in heat transfer:

1. Prescribed surface concentration

2. Prescribed mass flux at a surface

3. Mass flux transferred by fluid convection

$$N - k_m (C - C_b) \tag{20}$$

where k_m is a mass transfer coefficient and C_b is the concentration in the bulk fluid stream.

C. Fluid Velocity in the Melt

Fluid flow in the melt is natural or forced. Natural convection can occur due to density variations caused by temperature and solute concentration variations in the melt. Flow due to surface tension variations (the Marangoni flow) can also occur if the melt free surface is free of adsorbed layers and has temperature or solute concentration variations. Forced convection may be caused by external force such as crystal rotation, crucible rotation, etc. If the melt is heated by low-frequency induction heating, flow may be caused by the electromagnetic forces.

These flows are satisfactorily expressed by the Navier-Stokes equation [5]:

$$\left(\frac{\partial \vec{v}}{\partial t} + \vec{v} \cdot \nabla \vec{v} \right) = \vec{X} - \nabla p + \nabla(\mu \, \nabla \vec{v}) \tag{21}$$

where $\underline{X}$ is the body force acting on the fluid element. The buoyant force due to a density variation is given by [5]

$$\vec{X} = -[\beta \rho (T - \bar{T}) + \beta_m \rho (C - \bar{C})]\vec{g} \tag{22}$$

where β is the coefficient of thermal expansion $-(\partial \ln \rho / \partial T)_C$, β_m is the concentration coefficient of the volumetric expansion $-(\partial \ln \rho / \partial C)_T$, $\bar{T}$ is the average temperature, $\bar{C}$ is the average concentration, and $\vec{g}$ is the acceleration vector due to gravity.

Conservation of mass requires that the continuity equation hold, i.e.,

$$\frac{\partial \rho}{\partial t} + \nabla \cdot (\rho \vec{v}) = 0 \tag{23}$$

The velocity at the boundary is specified.

1. At the fluid-solid boundary, the condition of no slip must be satisfied:

$$\vec{v} = \vec{v}_s \tag{24}$$

where $\vec{v}_s$ is the velocity of the solid.

2. At the nonmixing fluid-fluid boundary, the velocity is continuous and the forces acting one against the other are equal [8]:

$$\vec{v}_1 = \vec{v}_2 \tag{25}$$

$$\sum_{k=1}^{3} \sigma_{1,ik} n_k = \sum_{k=1}^{3} \sigma_{2,ik} n_k, \quad i = 1, 2, 3 \tag{26}$$

where $\vec{n} = (n_1, n_2, n_3)$ is the unit normal vector which directs from 2 to 1 media and σ_{ij} is the component of the stress tensor defined by

$$\sigma_{ik} = - p\delta_{ik} + \mu\left(\frac{\partial v_i}{\partial x_k} + \frac{\partial v_k}{\partial x_i}\right) \tag{27}$$

where $(x_1, x_2, x_3) = (x, y, z)$ and $(v_1, v_2, v_3) = (v_x, v_y, v_z)$. If the surface tension is effective at the boundary, Eq. (26) is replaced by

$$\sum_{k=1}^{3} (\sigma_{1,ik} - \sigma_{2,ik}) n_k = \gamma K n_i + \frac{\partial \gamma}{\partial x_i}, \quad i = 1, 2, 3 \tag{28}$$

where K is the curvature of the surface and γ is the surface tension.

3. At the free surface, the boundary condition is given from Eq. (28) by

$$\sum_{k=1}^{3} \sigma_{ik} n_k = \gamma K n_i + \frac{\partial \gamma}{\partial x_i}, \quad i = 1, 2, 3 \tag{29}$$

D. Interface Shape

If the crystal is neither growing nor melting, the interface shape is given by the melting point isotherm. The melting point depends on the solute concentration and is determined by the phase diagram. For a dilute binary solution it may be given by

$$T_{eq}(C) = T_m + mC \tag{30}$$

where T_m is the melting point of the pure solvent and m is the slope of the liquidus line.

For a curved surface, the equilibrium temperature is given by [9]

$$T_{eq}(C, K) = T_{eq}(C) - \frac{\gamma}{\Delta S} K \tag{31}$$

where γ is the interfacial energy, ΔS is the entropy of fusion, and K is the curvature of the interface.

For crystal growth, supercooling is required to provide the driving force for the growth. The interface may no longer be isothermal, but must be a nonisothermal surface reflecting anisotropy of the crystal lattice and growth kinetics. Then additional conditions (response functions) are needed to specify the interface position and the interfacial growth velocity [10]. For a binary system, two response functions, the growth velocity normal to the surface f and the solute concentration in the solid C_s, may be expressed in terms of given interfacial supercooling ΔT and solute concentration in the liquid C_ℓ:

$$f = f(\Delta T, C_\ell), \quad C_s = C_s(\Delta T, C_\ell) \tag{32}$$

If the interface is at local equilibrium,

$$\Delta T = 0, \quad C_{eq,s} = k_0 C_{eq,\ell} \tag{33}$$

where $C_{eq,i}$ is the equilibrium solute concentration and k_0 is the equilibrium segregation coefficient. Even though the equilibrium shape of the crystal may be faceted, the faceting during growth is usually a result of the growth kinetics and the assumption of local equilibrium at the interface no longer holds.

For a pure system,

$$f = f(\Delta T) \tag{34}$$

The generalized growth laws for different known mechanisms are summarized in Appendix C.

E. Basic Assumptions

In previous sections, the equations governing heat transfer during the crystal growth are given. Generally, these equations cannot be solved analytically. It is also difficult to obtain exact numerical solutions. Therefore, approximate solutions, which still have the essential features of the crystal growth, are sometimes obtained by making many simplifying assumptions. These assumptions are summarized as follows:

1. The temperature at the interface is melting-point, i.e., the growth is controlled by heat transfer.
2. The heat flow is at steady state.
3. In the crystal, heat is transferred by conduction only.
4. In the melt, the heat is transferred both by conduction and by convection.
5. There is no heat generation in the crystal or in the melt.
6. The heat is dissipated from the crystal and melt free surfaces by gaseous convection or by radiation.
7. Latent heat is liberated at the interface.
8. Crystal diameter is constant.
9. The crucible temperature is constant.
10. Capillary rise of the melt is neglected and a planar meniscus is assumed.
11. All of the physical properties are constant, isotropic, and independent of temperature.

Assumptions 2 and 3 are satisfactorily applied if the Péclet number is small. The validity of some assumptions is discussed in Sec. IX.

V. HEAT FLOW IN THE CRYSTAL

A. One-dimensional Model

The one-dimensional model is very useful for a rough understanding of heat transfer during solidification. An analytical solution is usually obtainable. In this model, the radial temperature gradient is neglected and a planar interface is assumed. A heat balance over a differential element in a cylinder gives

$$\rho_s c_{ps}\left(\frac{\partial T}{\partial t} + f\frac{\partial T}{\partial z}\right) = \frac{\partial}{\partial z}\left(k_s \frac{\partial T}{\partial z}\right) - \frac{2}{a} q_{dis} \qquad (35)$$

where q_{dis} is the heat dissipation rate from the crystal surface and z is the axial distance from the interface.

This equation is often applicable to noncircular crystals with slight change so that the radius a is replaced by the effective radius defined by

$$a_{eff} = \frac{2S}{L_p} \qquad (36)$$

where S is the cross-sectional area and L_p is the periphery of the crystal.

Before discussing the validity of the one-dimensional model, it is recommended that the temperature distribution in the crystal be measured directly if possible. From the temperature distribution, one can deduce the heat dissipation from the crystal surface by

$$q_{dis} = \frac{a}{2} \frac{d}{dz}\left(k_s \frac{dT}{dz}\right) \qquad (37)$$

which is obtained from Eq. (35) if the Péclet number is small.

In Table 3, the typical values of the thermal properties are shown for some Czochralski growth materials.

1. Linear Heat Dissipation: $q_{dis} = h_a(T - T_a)$
Kuo and Wilcox [11] investigated this case in detail. Dimensionless variables are introduced:

$$\theta = \frac{T - T_a}{T_m - T_a}, \quad Z = \frac{z}{a}, \quad \tau = \frac{ft}{a} \qquad (38)$$

where T_a is the temperature of surroundings and T_m is the melting point. The crystal length is obtained by $\ell = ft$ and then the dimensionless crystal length is given by $L = ft/a = \tau$. With these dimensionless variables Eq. (35) for a constant thermal conductivity becomes

$$\frac{\partial^2 \theta}{\partial z^2} - \alpha \frac{\partial \theta}{\partial Z} - 2H\theta = \alpha \frac{\partial \theta}{\partial \tau} \qquad (39)$$

TABLE 3

Heat Transfer Properties for Various Materials

	Cu	W	Ge	Si	SBN	Cr_2O_3
Melting point T_m (K)	1356	3653	1210	1685	1470	2538
Thermal conductivity k_s (W cm^{-1} K^{-1})	3.3	0.895	0.174	0.220	0.008	0.04
Emissivity ε_s	0.056	0.36	0.59	0.46	0.88	0.8
Crystal radius[a] a(cm)	1	1	1	1	1	1
Biot number $H = \dfrac{\varepsilon_s \sigma T_m^3 a}{k_s}$	0.000240	0.111	0.0341	0.0567	1.98	1.85
Interfacial temperature gradient						
$\beta(0) = (0.8H)^{1/2}$	0.0139	0.298	0.165	0.213	1.26	1.22
$G_s = \dfrac{T_m}{a} \beta(0)$ (K cm^{-1})	18.8	1090	200	359	1850	3090
Characteristic crystal length						
$z^c = (0.8H)^{-1/2}$	72.7	4.53	6.06	4.70	0.794	0.821
$z^c = aZ^c$ (cm)	72.2	4.53	6.06	4.70	0.794	0.821

[a]Assumed.
Sources: Refs. 1, 71.

where $H = h_a a/k_s$ is the Biot number and $\alpha = f\rho_s c_{ps} a/k_s$ is the Péclet number. The Biot number represents roughly the ratio of heat dissipation from the crystal surface to axial conduction in the crystal. The Péclet number represents roughly the ratio of the heat carried by the movement of the crystal to that conducted in the crystal. If the crystal growth rate is slow, the Biot number is the prime factor which characterizes the heat transfer in the crystal.

The boundary conditions are:

$$\theta = 1 \text{ at the interface } (Z = 0)$$

$$\frac{\partial \theta}{\partial Z} = -H_e \theta \text{ at the cold end } (Z = L) \tag{40}$$

where $H_e = h_e a/k_s$. If the cold end is thermally insulated, $h_e = 0$, or, if perfectly cooled, $h_e = \infty$. When a Czochralski crystal with a neck and shoulder is growing, the boundary condition at the cold end may be regarded as $h_e = h_a$.

a. SMALL PÉCLET NUMBER

The Péclet number is small for slow growth and the terms containing it in Eq. (39) may be neglected:

$$\frac{\partial^2 \theta}{\partial Z^2} = 2H\theta \tag{41}$$

Equation (41) may easily be solved to yield:

$$\theta = \frac{(H_e/H)\sinh(2H)^{1/2}(L-Z) + (2/H)^{1/2}\cosh(2H)^{1/2}(L-Z)}{(H_e/H)\sinh(2H)^{1/2}L + (2/H)^{1/2}\cosh(2H)^{1/2}L}$$

The temperature gradient at the interface is given by

$$\beta(0) = -\left(\frac{d\theta}{dZ}\right)_{Z=0} = (2H)^{1/2}\frac{(H_e/H)\cosh(2H)^{1/2}L + (2/H)^{1/2}\sinh(2H)^{1/2}L}{(H_e/H)\sinh(2H)^{1/2}L + (2/H)^{1/2}\cosh(2H)^{1/2}L} \tag{43}$$

as shown in Fig. 2. If $H = H_e = 2$, the temperature gradient is independent of the crystal length.

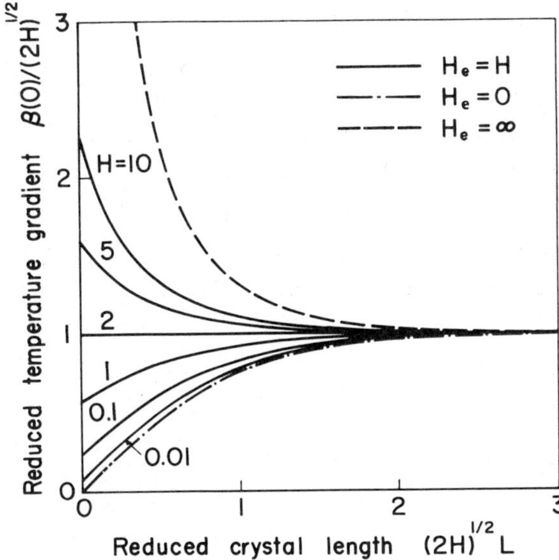

FIG. 2. Temperature gradient at the interface as a function of crystal length.

For a short crystal, the interfacial temperature gradient depends on the thermal condition at the cold end and it is approximately given by

$$\beta(0) = \begin{cases} (2H)L & \text{for } H_e = 0 \\ H[1 + (2 - H)L] & \text{for } H_e = H \\ \dfrac{1}{L} & \text{for } H_e = \infty \end{cases} \qquad (44)$$

For a long crystal, the temperature and interfacial temperature gradient are given respectively by

$$\theta = \exp[-(2H)^{1/2}z] \qquad (45)$$

and

$$\beta(0) = (2H)^{1/2} \qquad (46)$$

The interfacial temperature gradient is then independent of the thermal condition on the cold end. If $H_e = \infty$, the interfacial temperature

gradient monotonically decreases with crystal length and approaches
the value for a long crystal. If $H_e = 0$, it monotonically increases
with crystal length. If $H_e = H$, it behaves like the case with $H_e = 0$
for $H < 2$ and like the case with $H_e = \infty$ for $H > 2$.

The crystal length (or equivalently) time, when the interfacial
temperature gradient becomes constant is then estimated. It is defined by the cross-point between the asymptotic profiles of the interfacial temperature gradient for a short crystal and for a long crystal
and given by

$$
L^c = \frac{\ell^c}{a} =
\begin{cases}
(2H)^{-1/2} & \text{for } H_e = 0,\ \infty \\[2ex]
\dfrac{(2H)^{-1/2}}{1 + (H/2)^{1/2}} \leqq (2H)^{-1/2} & \text{for } H_e = H
\end{cases}
\tag{47}
$$

The critical length of Ge was $6a^{1/2}$ for $H = 0.015$ [17], which is consistent with Eq. (47).

$b.$ *FINITE PÉCLET NUMBER*

The heat carried by the movement of the crystal is now considered [11-13]. For simplicity, a semiinfinite rod at steady state is
assumed. The temperature and the interfacial temperature gradient
respectively are given by

$$
\theta = \exp\{\tfrac{1}{2}[\alpha - (\alpha^2 + 8H)^{1/2}]z\}
\tag{48}
$$

and

$$
\beta(0) = \tfrac{1}{2}[(\alpha^2 + 8H)^{1/2} - \alpha]
\tag{49}
$$

For a slow growth rate ($\alpha^2 \ll 8H$), Eqs. (48) and (49) becomes Eqs.
(45) and (46).

$2.$ *Radiative Heat Dissipation:* $q_{dis} = \varepsilon_s \sigma (T^4 - T_a^4)$
For high-temperature crystal growth, the heat is mainly dissipated
by radiation. This model is often used [1] and reduced to the previous model

$$h_a = \frac{\varepsilon_s \sigma (T^4 - T_a^4)}{T - T_a} \simeq 4\varepsilon_s \sigma T_a^3 \tag{50}$$

if $T \simeq T_a$. With the dimensionless variables

$$\theta = \frac{T}{T_m}, \quad \theta_a = \frac{T_a}{T_m}, \quad Z = \frac{z}{a}, \quad \tau = \frac{ft}{a} \tag{51}$$

Eq. (35) becomes

$$\frac{\partial^2 \theta}{\partial Z^2} - \alpha \frac{\partial \theta}{\partial Z} - 2H(\theta^4 - \theta_a^4) = \alpha \frac{\partial \theta}{\partial \tau} \tag{52}$$

where $H = \varepsilon_s \sigma T_m^3 a/k_s$ is the Biot number when radiation is dominant and $\alpha = f\rho_s c_{ps} a/k_s$ is the Péclet number. If the Péclet number is small, Eq. (52) becomes

$$\frac{\partial^2 \theta}{\partial Z^2} = 2H(\theta^4 - \theta_a^4) \tag{53}$$

a.   COOLED COLD END

If the heat flux at the cold end is also specified, the boundary conditions at the cold end are given by

$$\theta = \theta_a \quad \text{and} \quad \frac{d\theta}{dZ} = -\beta_e \quad \text{at the cold end } (Z = L) \tag{54}$$

The temperature is implicitly determined by the integral equation for $\theta_a = \theta$ [1]:

$$Z = \int_\theta^1 \frac{d\theta}{(0.8H\theta^5 + \beta_e^2)^{1/2}} \tag{55}$$

The interfacial temperature gradient is

$$\beta(0) = (0.8H + \beta_e^2)^{1/2} = \begin{cases} \beta_e & \text{for } 0.8H \ll \beta_e \\ (0.8H)^{1/2} & \text{for } 0.8H \gg \beta_e \end{cases} \tag{56}$$

If $0.8H < \beta$, the temperature is almost entirely determined by the conduction to the cold end. If $0.8H > \beta_e$, there is a position Z^c and the

corresponding temperature θ^C at which the dominant heat transfer mechanism changes. Near the interface $(\theta > \theta^C)$ the heat is dissipated predominantly by radiation from the crystal surface, while near the cold end $(\theta < \theta^C)$ the heat is dissipated by conduction to the cold end. The temperature gradient in the crystal is given by

$$\beta(Z) = (0.8H\theta^5 + \beta_e^2)^{1/2} = \begin{cases} \beta_e & \text{for } \theta < \theta^C \\ (0.8H\theta^5)^{1/2} & \text{for } \theta > \theta^C \end{cases} \tag{57}$$

Then the crossover temperature is given by

$$\theta^C = \left(\frac{\beta_e^2}{0.8H}\right)^{1/5} \tag{58}$$

The distance of the crossover temperature from the cold end is

$$z^C = \frac{\theta^C}{\beta_e} = [0.8H(\theta^C)^3 - 1/2] \geqq (0.8H)^{-1/2} \tag{59}$$

If the crystal length exceeds the critical length z^C, the interfacial temperature gradient becomes constant and is given by

$$\beta(0) = (0.8H)^{1/2} \tag{60}$$

b. *RADIATED COLD END*

The heat is also dissipated by radiation at the cold end and the boundary condition is

$$\left(\frac{d\theta}{dZ}\right)_{Z=L} = -H(\theta_e^4 - \theta_a^4) \tag{61}$$

where θ_e is the dimensionless temperature at the cold end. This model is applicable to a crystal with a neck and a shoulder [14]. The temperature is implicitly given by the integral equation for $\theta_a = 0$:

$$z(\theta) = \int_\theta^1 \frac{d\theta}{[0.8H(\theta^5 - \theta_e^5) + H^2\theta_e^8]^{1/2}} \tag{62}$$

and the interfacial temperature gradient is

$$\beta(0) = [0.8H(1 - \theta_e^5) + H^2\theta_e^8]^{1/2} \tag{63}$$

If the crystal is short, $\theta_e \simeq 1$, while if the crystal is long, $\theta_e \simeq 0$.   Therefore, the interfacial temperature gradient is approximated by

$$\beta(0) = \begin{cases} H\left[1 - 4\left(1 - \dfrac{1}{2H}\right)(1 - \theta_e)\right] \simeq H & \text{for a short crystal } (\theta_e \simeq 1) \\[4mm] (0.8H)^{1/2} & \text{for a long crystal } (\theta_e \simeq 0) \end{cases}$$

$$(64)$$

The distance of the cold end from the interface is given by

$$L = \frac{\ell}{a} = Z(\theta_e) \tag{65}$$

By eliminating θ_e from Eqs. (63) and (65), the relation between the interfacial temperature gradient and the crystal length is obtained. In Fig. 3, the interfacial temperature gradient is shown as a function of the crystal length for various Biot numbers.   The profile of the temperature gradient may be divided into three classes by the Biot number.   If $H > 0.8$, initially the temperature gradient rapidly decreases with crystal length and then gradually increases up to the constant level $(0.8H)^{1/2}$.   If $0.8 > H > 0.5$, it behaves similarly to the previous case $(0.8 < H)$ except its initial value is below the constant level.   And finally if $H < 0.5$, it monotonically increases up to the constant level.

The crystal length ℓ^c when the interfacial temperature gradient becomes constant is estimated.   It is defined by the cross-point between the asymptotic profiles of the temperature gradient for a short crystal and for a long crystal:

$$L^c = \frac{\ell^c}{a} = \begin{cases} \dfrac{1 - (0.8/H)^{1/2}}{4H} & \text{for } H > 0.8 \\[4mm] 0.4(0.8H)^{-1/2} - 0.5 & \text{for } H < 0.5 \end{cases}$$

$$(66)$$

This result may be compared with Eq. (59).

   c.   COLD END AT TEMPERATURE OF SURROUNDINGS

The temperature at the cold end is fixed at the temperature of the surroundings:

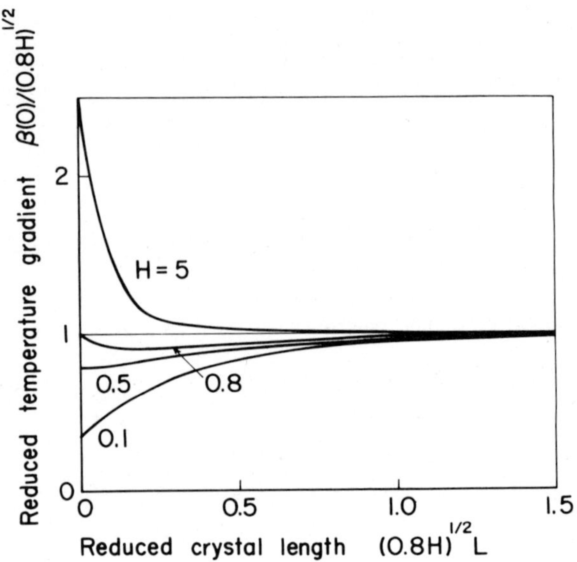

FIG. 3. Temperature gradient at the interface as a function of crystal length.

$$\theta \to \theta_a \quad \text{as} \quad Z \to \infty \tag{67}$$

If the thermal conductivity is independent of temperature, the temperature is implicitly given by

$$Z(\theta) = \int_\theta^1 \frac{d\theta}{[0.8H(\theta^5 - \theta_a^5)]^{1/2}} \tag{68}$$

and the interfacial temperature gradient is

$$\beta(0) = [0.8H(1 - \theta_a^5)]^{1/2} \tag{69}$$

If the thermal conductivity depends on temperature as $k_s = k_m T_m/T$, the temperature is implicitly given by

$$Z(\theta) = \int_\theta^1 \frac{d\theta/\theta}{[H\{\theta^4 - \theta_a^4(1 \pm 4 \ln \theta/\theta_a)\}]^{1/2}} \tag{70}$$

and the interfacial temperature gradient is given by

$$\beta(0) = [H\{1 - \theta_a^4(1 - 4\ln\theta_a)\}]^{1/2} \tag{71}$$

This model is sometimes used for semiconductor crystals [15].

If the thermal conductivity varies with temperature as $k_s = k_m T/T_m$, the temperature is implicitly

$$Z(\theta) = \int_\theta^1 \frac{\theta\, d\theta}{[(2H/3)(\theta^6 - 3\theta_a^3\theta^3 + 2\theta_a^6)]^{1/2}} \tag{72}$$

and the interfacial temperature gradient is

$$\beta(0) = \left[\frac{2H}{3}(1 - 3\theta_a^3 + 2\theta_a^6)\right]^{1/2} \tag{73}$$

This model is sometimes used for metals [16].

When $\theta_a = 0$, Eqs. (68), (70), and (72) are easily integrated to yield $\theta = [1 + (1.8H)^{1/2}Z]^{-2/3}$, $\theta = (1 + 2H^{1/2}Z)^{-1/2}$, and $\theta = [1 + (2H/3)^{1/2}Z]^{-1}$ and Eqs. (69), (71), and (73) become $\beta(0) = (0.8H)^{1/2}$, $\beta(0) = H^{1/2}$, and $\beta(0) = (2H/3)^{1/2}$, respectively.

This model may be applicable to the case with an insulated cold end whose temperature is θ_e in place of θ_a. The temperature at the cold end is implicitly determined from the crystal length by

$$L = \frac{\ell}{a} = Z(\theta_e) \tag{74}$$

B. Three-dimensional Model

The one-dimensional model is very useful for a rough estimate of the heat transfer in the crystal, but does not give correct temperature field because the radial variation of temperature is neglected. Especially for high-temperature crystal growth a steep radial temperature gradient is easily induced and the one-dimensional model is no longer valid because the Biot number is usually large [12].

The basic equation governing the temperature is already given by Eq. (6) for isotropic crystal. For a constant thermal conductivity, Eq. (6) in cylindrical coordinates becomes

$$\rho_s c_{ps}\left(\frac{\partial T}{\partial t} + f\,\frac{\partial T}{\partial z}\right) = k_s\left[\frac{1}{r}\frac{\partial}{\partial r}\left(r\,\frac{\partial T}{\partial r}\right) + \frac{\partial^2 T}{\partial z^2}\right] \tag{75}$$

where f is the growth rate. With dimensionless variables

$$\theta = \frac{T - T_a}{T_m - T_a},\quad R = \frac{r}{a},\quad Z = \frac{z}{a},\quad \tau = \frac{ft}{a} \tag{76}$$

Eq. (75) becomes

$$\alpha\left(\frac{\partial \theta}{\partial \tau} + \frac{\partial \theta}{\partial Z}\right) = \frac{1}{R}\frac{\partial}{\partial R}\left(R\,\frac{\partial \theta}{\partial R}\right) + \frac{\partial^2 \theta}{\partial Z^2} \tag{77}$$

where $\alpha = f\rho_s c_{ps} a/k_s$ is the Péclet number. For very low Péclet number, Eq. (77) becomes the Laplace equation:

$$\Delta\theta = \frac{1}{R}\frac{\partial}{\partial R}\left(R\,\frac{\partial \theta}{\partial R}\right) + \frac{\partial^2 \theta}{\partial Z^2} = 0 \tag{78}$$

If the heat fluxes from the crystal surface and from the cold end are given by $h_a(T - T_a)$ and $h_e(T - T_a)$, respectively, boundary conditions for a finite crystal are

$$\left(\frac{\partial T}{\partial r}\right) = 0 \quad \text{at the axis } (r = 0)$$

$$T = T_m \quad \text{at the interface } (z = 0)$$

$$-k_s\left(\frac{\partial T}{\partial r}\right) = h_a(T - T_a) \quad \text{at the crystal surface } (r = a)$$

$$-k_s\left(\frac{\partial T}{\partial z}\right) = h_e(T - T_a) \quad \text{at the cold end } (z = \ell) \tag{79}$$

Then, the boundary conditions in dimensionless form are

$$\left(\frac{\partial \theta}{\partial R}\right) = 0 \quad \text{at } R = 0$$

$$\theta = 1 \quad \text{at } Z = 0$$

$$\left(\frac{\partial \theta}{\partial R}\right) = -H\theta \quad \text{at } R = 1$$

$$\left(\frac{\partial \theta}{\partial Z}\right) = -H_e\,\theta \quad \text{at } Z = L \tag{80}$$

where $H = h_a a/k_s$ is the Biot number, $H_e = h_e a/k_s$, and $L = \ell/a$.

The temperature is given by [17]

$$\theta = 2H \sum_{n=1}^{\infty} \frac{1}{H^2 + \alpha_n^2} \frac{J_0(\alpha_n R)}{J_0(\alpha_n)} \frac{\alpha_n \cosh \alpha_n (L - Z) + H \sinh \alpha_n (L - Z)}{\alpha_n \cosh \alpha_n L + H \sinh \alpha_n L} \tag{81}$$

where α_n is the nth root of $\alpha_n J_1(\alpha_n) - H J_0(\alpha_n) = 0$. For a long crystal, Eq. (82) becomes [18]

$$\theta = 2H \sum_{n=1}^{\infty} \frac{1}{H^2 + \alpha_n^2} \frac{J_0(\alpha_n R)}{J_0(\alpha_n)} \exp(-\alpha_n Z) \tag{82}$$

If $H \ll 1$, the temperature is approximately given by the first term in Eq. (81):

$$\theta = \frac{1 - HR^2/2}{1 - H/2} \exp\left[-(2H)^{1/2} Z\right] \tag{83}$$

Apart from the radial dependence, the axial temperature is about the same as Eq. (45) which is obtained by the one-dimensional model.

This model is applied for the growth of Ge [17] and $ZnWO_4$ [19]. The temperature distribution was well expressed by Eq. (81) with $H = 0.015$ and $T_a = 420°C$ for Ge [17]. For the growth of $ZnWO_4$ [19], the temperature distribution was well expressed by Eq. (83) with $H = 0.042$ and $T_a = 425°C$. However, T_a does not always signify the actual ambient gas temperature in the system. This suggests that the ambient temperature is dependent on the spacial position.

C. Summary

The effects of the various factors which may be used to control the heat transfer in the crystal are summarized below.

To increase the axial temperature gradient in the crystal:

1. The position of the heat source relative to the crucible is moved downward.

2. A large crucible is used.

3. A gas jet is blown on the crystal.

    To decrease the axial temperature gradient in the crystal:

4. An afterheater, reflector, or thermal shield is used.

VI. HEAT AND FLUID FLOW IN THE MELT

The melt plays an important role in crystal growth. The temperature
of the melt controls the heat transferred from the melt to the interface by its influence on the interfacial temperature gradient. The
crystal diameter may be controlled by the bulk melt temperature for
low-temperature growth, even if the temperature field in the melt is
not known. For high-temperature crystal growth, however, diameter
control is often very difficult because of the typically high Biot
number. Crystal rotation is useful to decrease thermal asymmetry
around the crystal by mixing the fluid. Crucible rotation also homo-
genizes the temperature in the melt and makes the radial temperature
gradient around the crystal more symmetric.

The flow in the melt controls the segregation of the solute at
the interface. Radial solute homogeneity is much improved by increasing the crystal rotation. Burton et al. [20] indirectly showed that
the flow caused by crystal rotation is approximately given by Cochran's
flow [21]. For other flows, the detailed properties are not well
known. Fluid flow is usually induced both by buoyancy and by crystal
rotation. This flow influences heat transfer and the interface shape
as well as mass transfer of solute. It is necessary to know something
about the fluid flow in order to understand crystal growth. A detailed review on mass transfer has been given in this series [7].

Usually, capillary rise of the melt is neglected. A planar meniscus and planar interface are assumed. Attempts have been made to
obtain the heat flow near the interface, with some investigating the
influence of forced convection. The basic equations governing the
temperature and fluid flow in cylindrical coordinate for a constant
thermal conductivity and an incompressible pure fluid are obtained
from Eqs. (9), (21), and (23):

$$\frac{\partial T}{\partial t} + u\,\frac{\partial T}{\partial r} + w\,\frac{\partial T}{\partial z} = \alpha_t\left[\frac{1}{r}\frac{\partial}{\partial r}\left(r\,\frac{\partial T}{\partial r}\right) + \frac{\partial^2 T}{\partial z^2}\right] + \frac{\nu}{c_p}\,\phi \tag{84}$$

$$\frac{\partial u}{\partial t} + u\,\frac{\partial u}{\partial r} + w\,\frac{\partial u}{\partial z} - \frac{v^2}{r} = -\frac{1}{\rho}\frac{\partial p}{\partial r} + \nu\left\{\frac{\partial}{\partial r}\left[\frac{1}{r}\frac{\partial}{\partial r}\,(ru)\right] + \frac{\partial^2 u}{\partial z^2}\right\} \tag{85}$$

$$\frac{\partial v}{\partial t} + u\,\frac{\partial v}{\partial r} + w\,\frac{\partial v}{\partial z} + \frac{uv}{r} = \nu\left\{\frac{\partial}{\partial r}\left[\frac{1}{r}\frac{\partial}{\partial r}\,(rv)\right] + \frac{\partial^2 v}{\partial z^2}\right\} \tag{86}$$

$$\frac{\partial w}{\partial t} + u\,\frac{\partial w}{\partial r} + w\,\frac{\partial w}{\partial z} = -\beta g\,(T - \overline{T}) - \frac{1}{\rho}\frac{\partial p}{\partial z} + \nu\left[\frac{1}{r}\frac{\partial}{\partial r}\left(r\,\frac{\partial w}{\partial r}\right) + \frac{\partial^2 w}{\partial z^2}\right] \tag{87}$$

$$\frac{\partial}{\partial r}\,(ru) + \frac{\partial}{\partial z}\,(rw) = 0 \tag{88}$$

where z is the axial distance from the interface. Dimensionless variables are defined with a characteristic length L_c, time Ω_c^{-1}, temperature ΔT, and relative velocity of a particular motion U_c:

$$R = \frac{r}{L_c}, \quad Z = \frac{z}{L_c}, \quad \tau = \Omega_c t$$

$$\theta = \frac{T - T_m}{\Delta T}, \quad U = \frac{u}{U_c}, \quad V = \frac{v}{U_c}, \quad W = \frac{w}{U_c}, \quad P = \frac{p}{\rho\Omega_c L_c U_c} \tag{89}$$

With these dimensionless variables, Eqs. (84)-(88) become

$$\frac{1}{Ro}\frac{\partial\theta}{\partial\tau} + U\,\frac{\partial\theta}{\partial R} + W\,\frac{\partial\theta}{\partial Z} = \frac{1}{PrRe}\left[\frac{1}{R}\frac{\partial}{\partial R}\left(R\,\frac{\partial}{\partial R}\right) + \frac{\partial^2\theta}{\partial Z^2}\right] + \frac{1}{N_\mu}\,\phi \tag{90}$$

$$\frac{1}{Ro}\frac{\partial U}{\partial\tau} + U\,\frac{\partial U}{\partial R} + W\,\frac{\partial U}{\partial Z} - \frac{V^2}{R} = -\frac{\partial P}{\partial R} + \frac{1}{Re}\left\{\frac{\partial}{\partial R}\left[\frac{1}{R}\frac{\partial}{\partial R}\,(RU)\right] + \frac{\partial^2 U}{\partial Z^2}\right\} \tag{91}$$

$$\frac{1}{Ro}\frac{\partial V}{\partial\tau} + U\,\frac{\partial V}{\partial R} + W\,\frac{\partial V}{\partial Z} + \frac{UV}{R} = -\frac{1}{Re}\left\{\frac{\partial}{\partial R}\left[\frac{1}{R}\frac{\partial}{\partial R}\,(RV)\right] + \frac{\partial^2 V}{\partial Z^2}\right\} \tag{92}$$

$$\frac{1}{Ro}\frac{\partial W}{\partial\tau} + U\,\frac{\partial W}{\partial R} + W\,\frac{\partial W}{\partial Z} = -\frac{Gr}{Re^2}\,(\theta - \overline{\theta}) - \frac{\partial P}{\partial Z} + \frac{1}{Re}\left[\frac{1}{R}\frac{\partial}{\partial R}\left(R\,\frac{\partial W}{\partial R}\right) + \frac{\partial^2 W}{\partial Z^2}\right] \tag{93}$$

$$\frac{\partial}{\partial R}\,(RU) + \frac{\partial}{\partial Z}\,(RW) = 0 \tag{94}$$

where

$$\Phi = 2\left[\left(\frac{\partial U}{\partial R}\right)^2 + \left(\frac{U}{R}\right)^2 + \left(\frac{\partial W}{\partial Z}\right)^2\right] + \left(\frac{\partial V}{\partial R} - \frac{V}{R}\right)^2 + \left(\frac{\partial W}{\partial R} + \frac{\partial U}{\partial Z}\right)^2 + \left(\frac{\partial V}{\partial Z}\right)^2$$

$$(95)$$

$Ro = L_c \Omega_c / U_c$ is the Rossby number, $Pr = \nu/\alpha_t$ is the Prandtl number, $Re = L_c U_c / \nu$ is the Reynolds number, $N_\mu = \nu(U_c/L_c)/(c_p \Delta T)$ is a diffusivity coefficient, $Gr = g\beta \Delta T L_c^3/\nu^2$ is the Grashof number, and $\bar{\theta} = (\bar{T} - T_m)/\Delta T$. The scaling factors $(L_c, \Omega_c^{-1}, \Delta T, U_c)$ are variously defined.

For numerical convenience, a vorticity function χ and a stream function ψ are introduced, which are defined respectively by

$$\chi = \frac{\partial U}{\partial Z} - \frac{\partial W}{\partial R}$$

$$(96)$$

and

$$U = \frac{1}{R}\frac{\partial \psi}{\partial Z}, \quad W = -\frac{1}{R}\frac{\partial \psi}{\partial R}$$

$$(97)$$

By cross-differentiating Eqs. (91) and (93) and eliminating the pressure terms, one obtains a vorticity transport equation:

$$\frac{1}{Ro}\frac{\partial \chi}{\partial \tau} + U\frac{\partial \chi}{\partial R} + W\frac{\partial \chi}{\partial Z} - \frac{U\chi}{R} - \frac{\partial}{\partial Z}\frac{V^2}{R} = \frac{Gr}{Re^2}\frac{\partial \theta}{\partial R} + \frac{1}{Re}\left\{\frac{\partial}{\partial R}\left[\frac{1}{R}\frac{\partial}{\partial R}(R\chi)\right]\right.$$

$$\left. + \frac{\partial^2 \chi}{\partial Z^2}\right\}$$

$$(98)$$

The stream function automatically satisfies with the continuity equation, Eq. (94). The equation governing the stream function is obtained by substituting Eq. (97) into Eq. (96):

$$\frac{\partial^2 \psi}{\partial R^2} - \frac{1}{R}\frac{\partial \psi}{\partial R} + \frac{\partial^2 \psi}{\partial Z^2} = R\chi$$

$$(99)$$

Then the basic equations governing the flow in the melt are now given by Eqs. (92), (98), and (99) and the flow in the bulk melt is characterized by four dimensionless numbers, the Prandtl number, the Reynolds number, the diffusivity coefficient, and the Grashof number.

Usually, the diffusivity coefficient is large and the last term in
Eq. (90) is satisfactorily neglected.

A. Boundary Layer Theory

The interface is assumed to be an infinite rotating disk, with the
rotation causing forced convection in the melt. The crucible is
also assumed to be an infinite rotating disk. For fluid flow near
the interface, boundary layer theory may be applied as long as the
Reynolds number exceeds $\sim$100 [22]. If the flow is pure forced convection, the heat and mass flow may be expressed by the same type of
equation. When the boundary conditions also have the same forms, one
obtains solutions by analogy. The details of analogy between the heat
and mass transfer are given by Wilcox [7]. The solution of a heat
transfer problem may be obtained from a mass transfer problem by re-
placement of the solute concentration by the temperature and of the
Schmidt number $Sc = \nu/D$ (D is the diffusion coefficient) by Prandtl
number and vice versa.

1. *Steady State Flow*

The flow between two rotating disks may be divided into three classes
with the ratio of the rotation rates between the crystal and the cru-
cible [23], $s = \omega_c/\omega_s$, where ω_s is the crystal rotation rate and ω_c
is the crucible rotation rate, as shown in Fig. 4. If $s > 1$, there
is radial flow inward near the interface and an axial flow away from
the interface. If $1 > s \geq 0$, the axial velocity is towards the interface. The interface then acts as a centrifugal fan, throwing the
fluid radially outward and drawing it axially inward. Finally, if
$s < 0$, the interface and the crucible rotate in an opposite direction
and there is a plane which divides the flow into two self-confined
regions. Especially, if $s = 1$, the fluid rotates around the axis as
if it were a solid body.

The motion of the disk is communicated to the fluid through a
viscous boundary layer of thickness $\delta = (\nu/\Omega_c)^{1/2}$. Within this layer
(the Ekman layer), the Coriolis force is balanced by the viscous shear
and the pressure is a constant. The flow within the Ekman layer is
sometimes called the Ekman flow.

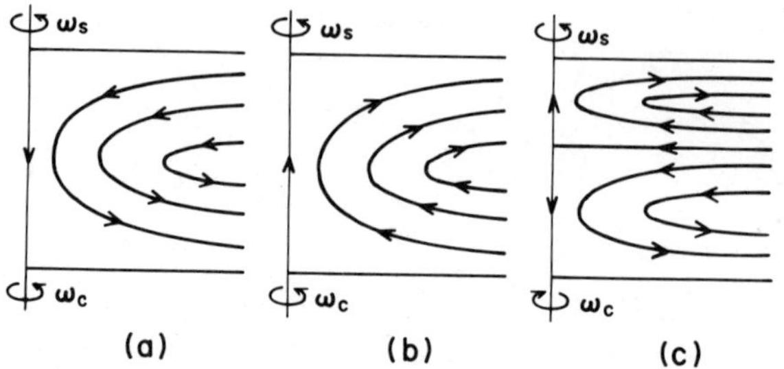

FIG. 4. Flow patterns caused by two infinite rotating disks: (a)
$s = \omega_c/\omega_s > 1$; (b) $1 > s \geq 0$; (c) $s < 0$. (From Ref. 23; reprinted
with permission.)

a. FLOW IN A DEEP MELT

Usually, the scaling factors $[(\nu/\omega_s)^{1/2}, \omega_s^{-1}, T_c - T_m, (\nu\omega_s)^{1/2}]$
are used for a fluid with a depth $d \gg (\nu/\omega_s)^{1/2}$, where ν is the kinematic viscosity, ω_s is the crystal rotation rate, T_c is the crucible
temperature, and T_m is the melting point. If the ratio Gr/Re^2 is
small, the buoyancy term in Eq. (98) is safely neglected and then
the flow is almost pure forced convection.

The boundary conditions are

$$T = T_m \quad u = 0 \quad v = r\omega_s \quad w = 0 \quad \text{at the interface } (z = 0)$$

$$T = T_c \quad u = 0 \quad v = r\omega_c \qquad \text{at the other end } (z \to \infty) \qquad (100)$$

The following assumptions are made for the velocity components and
pressure [24]:

$$U = RF(\zeta) \tag{101}$$

$$V = RG(\zeta) \tag{102}$$

$$W = H(\zeta) \tag{103}$$

$$P = P(\zeta) \tag{104}$$

where $\zeta = z(\omega_s/\nu)^{1/2}$ is used in place of Z. Inserting these equations
into Eqs. (90)-(94), one obtains a system of five simultaneous ordi-
nary equations for the functions θ, F, G, H, and P:

$$\theta'' - PrH\theta' = 0 \tag{105}$$

$$F'' - F'H + G^2 - F^2 - s^2 = 0 \tag{106}$$

$$G'' - G'H - 2FG = 0 \tag{107}$$

$$H'' - H'H - P' = 0 \tag{108}$$

$$H' + 2F = 0 \tag{109}$$

where the prime denotes the differentiation on ζ. The boundary conditions are obtained from Eq. (100):

$$\theta = 0, \quad U = 0, \quad V = 1, \quad W = 0 \quad \text{at } \zeta = 0$$

$$\theta = 1, \quad U = 0, \quad V = s, \qquad \text{at } \zeta \to \infty \tag{110}$$

Equation (105) is easily integrated and the temperature and the interfacial temperature gradient are given respectively by [24]

$$\theta = \frac{1}{\Delta_t} \int_0^\zeta \exp\left[\Pr \int_0^{\zeta_1} H(\zeta_2) \ d\zeta_2 \right] d\zeta_1 \tag{111}$$

$$\theta' = \frac{1}{\Delta_t} \tag{112}$$

where

$$\Delta_t - \delta_t \left(\frac{\omega_s}{\nu}\right)^{1/2} = \int_0^\infty \exp\left[\Pr \int_0^\zeta H(\zeta_1) \ d\zeta_1 \right] d\zeta \tag{113}$$

and δ_t is sometimes called the thermal layer thickness [24]. It is constant over an infinite rotating disk. The asymptotic expression for $\theta'(0)$ is given by [25]

$$\theta'(0) = \begin{cases} H(\infty)\Pr & \text{for } \Pr \ll 1 \\[2ex] \dfrac{[-H''(0)/6]^{1/3}\Pr^{1/3}}{T(4/3)} & \text{for } \Pr \gg 1 \end{cases} \tag{114}$$

The exact solution for $s = 0$ is numerically given by Cochran [21] and Eq. (115) is explicitly given by

$$\theta'(0) = \begin{cases} 0.8844\Pr & \text{for } \Pr \ll 1 \\[2ex] 0.6204\Pr^{1/3} & \text{for } \Pr \gg 1 \end{cases} \tag{115}$$

In Fig. 5, $\theta'(0)$ and Δ_t are shown as functions of the Prandtl number.

Rogers and Lance [26] and Capper and Elwell [27] obtained the fluid flow for various s values. If $s > 1$, it is similar to Fig.

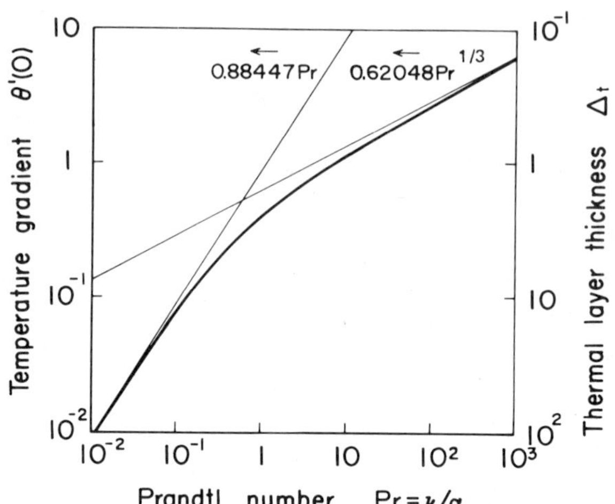

FIG. 5. Temperature gradient and thermal layer thickness as functions of Prandtl number.

4(a), and if $1 > s > -0.2$, the flow pattern is similar to Fig. 4(b). However, if $s < -0.2$, the solution is not obtained numerically because of the existence of an extra stagnation surface in the fluid as shown in Fig. 4(c). The flow pattern [27] within $0 > s > -0.2$ is different from that predicted by Batchelor [23] and the reason for the discrepancy is not yet explained. The results [26] are summarized in Table 4. For $1 > s > 0$, $\theta'(0)$ is asymptotically obtained from Eq. (115).

The above treatment is still valid when the thermal layer thickness is thin with respect to a finite crystal radius and when fluid flow is laminar. The former condition will not be satisfied if the Prandtl number is small, because a thick thermal layer is formed near the interface. The Reynolds number of the rotating disk, defined by $Re_r = a^2\omega_s/\nu$, is a measure of the state of the fluid flow [28]. If $Re_r < 2.4 \times 10^5$, the fluid flow is laminar and the above expression is valid. If $5.5 \times 10^5 > Re_r > 2.4 \times 10^5$, the flow begins to fluctuate and is no longer laminar. If $Re_r > 5.5 \times 10^5$, the flow is completely turbulent and the above expression is not valid. For a turbulent flow, the boundary layer thickness varies over the interface [28].

TABLE 4

Flow Properties of Infinite Rotating Disks

s	$-H''(0)/2$	$-H(\infty)$	1/s	$H''(0)/2$	$H(\infty)$
0.0	0.510233	0.88446	0.0	0.941971	1.36961
0.1	0.513397	0.91769	0.1	0.844923	1.19516
0.2	0.50187	0.86175	0.2	0.751682	1.03086
0.4	0.439580	0.65996	0.4	0.569080	0.72608
0.6	0.331470	0.43077	0.6	0.385194	0.45373
0.8	0.183467	0.20801	0.8	0.196121	0.21272
0.9	0.95936	0.10201	0.9	0.099014	0.10309
1.0	0.0	0.0	1.0	0.0	0.0

b. *FLOW IN A SHALLOW MELT*

A finite distance between the crystal and the crucible is considered in this model. Usually, the scaling factors $(d, \omega_s^{-1}, T_c - T_m, d\omega_s)$ are used. This flow is characterized by two Reynolds numbers, defined by $Re_d = d^2\omega_s/\nu$ and $Re_d' = d^2\omega_c/\nu$. The distance between the disks is normalized to unity. The flow pattern for small Reynolds numbers was obtained by Kreith and Viviand [29] for the ratio of the Reynolds number $s = Re_d'/Re_d = \omega_c/\omega_s$:

$$\frac{u}{d\omega_s} = \frac{(1-s)}{15} Re_d \rho \xi (1-\xi)[2(3+2s) - 5(3+s)(1-s)\xi$$
$$+ 5(1-s)\xi^2] \tag{117}$$

$$\frac{v}{d\omega_s} = \rho[1 - (1-s)\xi] \tag{118}$$

$$\frac{v}{d\omega_s} = -\frac{(1-s)}{30} Re_d \xi^2(1-\xi)^2[(3+s) - (1-s)\xi] \tag{119}$$

where $\rho = r/d$ and $\xi = z/d$. If $1 > s > -2/3$, the axial velocity is directed to the interface. If $-2/3 > s > -1$, the axial velocity changes sign once and the flow is divided into two self-confined regions by a dividing surface. The dividing surface is at the position

$$\xi_d = \frac{1}{2}\left[1 + 5\,\frac{1 + s}{1 - s}\right] \tag{120}$$

from the interface. The flow with $|s| > 1$ is just the flow with $1/s$ upside down.

Stewartson [22] studied the flow between two rotating disks experimentally and theoretically, for high Reynolds numbers. The bulk fluid rotates at an intermediate angular frequency when two disks are rotation in the same sense. If $s \simeq 1$ ($\omega_s \simeq \omega_c$), the flow near the interface is given by   22,30

$$\frac{u}{d\omega_s} = \frac{1 - s}{2}\,\rho\,\exp\left(-\mathrm{Re}_m^{1/2}\xi\right)\,\sin\left(\mathrm{Re}_m^{1/2}\xi\right) \tag{121}$$

$$\frac{v}{d\omega_s} = \rho\left[\frac{1 + s}{2} + \frac{1 - s}{2}\,\exp\left(-\mathrm{Re}_m^{1/2}\xi\right)\,\cos\left(\mathrm{Re}_m^{1/2}\xi\right)\right] \tag{122}$$

$$\frac{w}{d\omega_s} = -(1 - s)\,\mathrm{Re}_m^{1/2}\left\{1 - \exp\left(-\mathrm{Re}_m^{1/2}\xi\right)\left[\cos\left(\mathrm{Re}_m^{1/2}\xi\right)\right.\right.$$

$$\left.\left. + \sin\left(\mathrm{Re}_m^{1/2}\right)\right]\right\} \tag{123}$$

where $\mathrm{Re}_m = (1 + s)\,\mathrm{Re}_d/2$. If $s \simeq 0$, the main body is quiescent and the flow in the boundary layer is Cochran's flow [21]. If the disks rotate in an opposite sense, the main fluid does not rotate and a rapid change of flow occurs near the disks. The flow pattern approaches Cochran's flow. For $\mathrm{Re}_d = 100$, numerical solutions [31] have been obtained for various s values between zero and -1.

c. *FLOW WITH SUCTION*

Suction from the disk has been considered. An interfacial flow exists normal to the interface if the crystal is growing or melting. The boundary condition at the interface for an axial velocity is given by

$$w = -\left(\frac{\rho_s}{\rho_\ell}\right)f \tag{124}$$

where f is the growth rate. Now $f < 0$ if melting of the crystal occurs at the interface and $f > 0$ if growth of the crystal occurs.

When there is injection of fluid at the interface (f < 0), the fluid adjacent to the disk is near the same temperature as that of the disk and the temperature profile near the disk becomes flattened. The injected fluid effectively forms an insulating layer near the disk and the heat transfer to the disk decreases. On the other hand, when there is the suction of the fluid at the interface (f > 0), a large quantity of the bulk fluid is brought into the neighborhood of the disk. The temperature quickly drops near the disk, and so suction of the fluid increases heat transfer at the disk.

Sparrow and Gregg [32] numerically obtained the fluid flow by a single rotating disk. The boundary layer theory was applied and the boundary condition for a dimensionless axial velocity is given by using the same notation defined as in Eqs. (101)–(104):

$$H(0) = H_0 \tag{125}$$

where $H_0 = -(\rho_s/\rho_\ell) f/(\nu\omega_s)^{1/2}$. The interfacial temperature gradient $\theta'(0)$ is shown as a function of the fluid injection H_0 in Fig. 6. For large negative values of H_0, $\theta'(0)$ is given by

$$\theta'(0) = -H_0 Pr \tag{126}$$

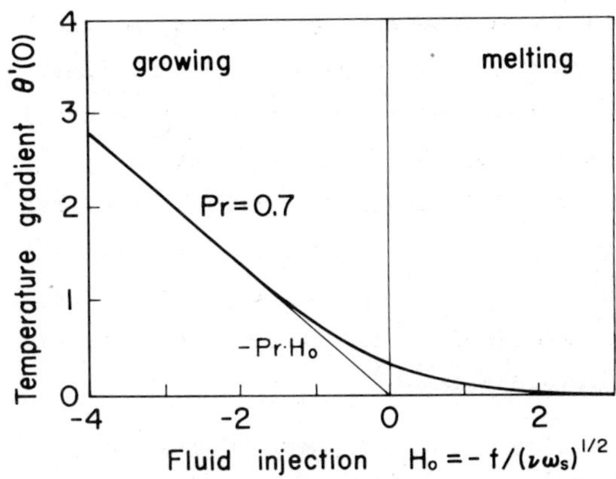

FIG. 6. Temperature gradient as a function of fluid injection for Pr = 0.7. [From E. M. Sparrow and J. L. Gregg, *J. Heat Transfer Trans. ASME, Ser. C, 82;* 299 (1960); reprinted with permission.]

Wilson [33] obtained a diffusion film thickness by analyzing the diffusion layer thickness obtained by Burton et al. [20]. By analogy, the interfacial temperature gradient is approximately given by

$$\theta'(0) = 0.62048 Pr^{2/3} (1 - H_0 Pr^{2/3}) \quad \text{for } 0.5 > -H_0 Pr^{2/3} > 0 \tag{127}$$

Wilson [34,35] also numerically obtained the flow between a fixed and rotating disks and showed that the interfacial temperature gradient is approximately given by

$$\theta'(0) = \frac{-H_0 Pr}{1 - \exp(-\Delta)} \quad \text{for } 1 \geq -H_0 Pr^{2/3} > 0 \tag{128}$$

where $\Delta = -1.86 H_0 Pr^{2/3}/(1 - 0.13 H_0 Pr^{2/3})$.

2. *Transient Flow*

The transient motion between two infinite rotating disks consists of three distinct phases [36]. The initial impulsive change in the angular velocity immediately produces a shear layer at each disk and a quasi-steady Ekman boundary layer develops from vorticity diffusion within a few revolutions, $\tau \simeq 1$. The Ekman layer acts as a sink for low-angular-momentum fluid in the interior, this fluid being replaced by high-angular-momentum fluid drawn from a larger radii. As the conditions in the interior approach the values appropriate to a final steady state, the Ekman layer decays. This happens in a dimensionless time of order $Re^{1/2}$. In the meantime, the small oscillations set up by the initial impulse have been modified very slightly in the interior and more markedly near the boundaries. They persist until they are finally destroyed by viscosity at a dimensionless time of the order Re. At this late time, the final steady state flow is attained. Three critical times are summarized below:

$\tau \simeq 1$ Dimensionless time for development of viscous boundary layer

$\tau \simeq Re^{1/2}$ Dimensionless time for spin-up

$\tau \simeq Re$ Dimensionless time for viscous decay of residual effects

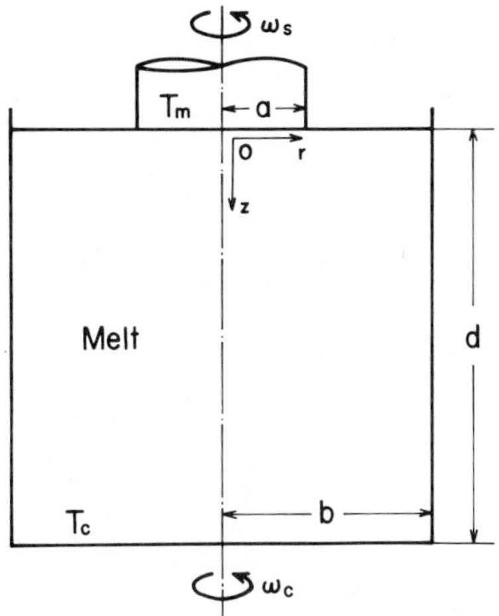

FIG. 7. Geometry of melt.

The time-dependent solution was obtained when one disk is impul-
sively started [31]. The time when the flow reaches a steady state
was obtained and the above general statements were confirmed.

B. Flow Simulation

In actual crystal growth, the flow in the crucible is normally not
observable, mainly because of the high temperature or optical opacity.
Then the flow has often been simulated experimentally by using a
transparent liquid with a dye [37-43]. This method has weak points
because the usual high-temperature difference, radiative heat transfer, and low Prandtl number are not easily modeled. However, this
method is advantageous because it is easy thereby to simulate the
flow in the crucible, except for the scaling problem.

On the other hand, differential equations have been solved directly by numerical technique [44-55], because it is almost impossible to obtain an analytical solution for an actual crystal growth
system. A crystal growth model [49] is shown in Fig. 7. A crystal
is growing from the melt in the cylindrical crucible. The interface
shape is assumed to be planar. The boundary conditions are:

At the interface $(0 < r \le a, z = 0)$:

$$T = T_m, \quad u = 0, \quad v = r\omega_s, \quad w = -\left(\frac{\rho_s}{\rho_\ell}\right) f_p \tag{129a}$$

At the melt free surface $(a < r < b, z = 0)$:

$$k_\ell \left(\frac{\partial T}{\partial z}\right) = \varepsilon_\ell \sigma (T^4 - T_a^4), \quad \mu \frac{\partial u}{\partial z} = -\frac{\partial \gamma}{\partial r}, \quad \frac{\partial v}{\partial z} = 0, \quad w = \frac{(\rho_s/\rho_\ell) f_p}{(b/a)^2 - 1}, \tag{129b}$$

At the crucible side $(r = b, 0 \le z \le d)$:

$$T = T_c, \quad u = 0, \quad v = b\omega_c, \quad w = 0 \tag{129c}$$

At the crucible bottom $(0 < r < b, z = d)$:

$$T = T_c, \quad u = 0, \quad v = r\omega_c, \quad w = 0 \tag{129d}$$

At the symmetric axis $(r = 0, 0 \le z \le d)$:

$$\frac{\partial T}{\partial r} = 0, \quad u = 0, \quad v = 0, \quad \frac{\partial w}{\partial r} = 0 \tag{129e}$$

where γ is the surface tension and f_p is the pulling rate. The second term in Eq. (129b) expresses the mechanical balance between the viscous force and the surface tension, which causes a flow called the Marangoni flow. The growth rate is not equal to the pulling rate because the melt drops during growth and it is given by

$$f = \frac{f_p}{1 - (a/b)^2} \tag{130}$$

Then the fourth term in Eq. (129b) expresses the falling rate of the melt height as the crystal grows.

With the scaling factors $(a, \omega_s^{-1}, T_c - T_m, a\omega_s)$, the boundary conditions including those of the vorticity and stream function are expressed by:

At the interface $(0 < R \le 1, Z = 0)$:

$$\theta = 0, \quad U = 0, \quad V = R, \quad W = -F_p, \quad \chi = \frac{\partial U}{\partial Z}, \quad \psi = \frac{R^2 F_p}{2} \tag{131a}$$

At the melt free surface $(1 < R < R_c, Z = 0)$:

$$\frac{\partial \theta}{\partial Z} = H_m \left[\left(1 + \frac{\Delta T}{T_m}\theta\right)^4 - \left(1 + \frac{\Delta T}{T_m}\theta_a\right)^4 \right], \quad \frac{\partial U}{\partial Z} = \frac{Ma_r}{PrRe_r}\frac{\partial \theta}{\partial R}, \quad \frac{\partial V}{\partial Z} = 0$$

$$W = \frac{F_p}{R_c^2 - 1}, \quad \chi = -\frac{\partial U}{\partial Z}, \quad \psi = \frac{(R_c^2 - R^2)F_p}{2(R_c^2 - 1)} \tag{131b}$$

At the crucible side $(R = R_c, 0 \leq Z \leq D)$:

$$\theta = 1, \quad U = 0, \quad V = sR_c, \quad W = 0, \quad \chi = -\frac{\partial W}{\partial R}, \quad \psi = 0 \tag{131c}$$

At the crucible bottom $(0 < R < R_c, Z = D)$:

$$\theta = 1, \quad U = 0, \quad V = sR, \quad W = 0, \quad \chi = \frac{\partial U}{\partial Z}, \quad \psi = 0 \tag{131d}$$

At the symmetric axis $(R = 0, 0 \leq Z \leq D)$:

$$\frac{\partial \theta}{\partial R} = 0, \quad U = 0, \quad V = 0, \quad \frac{\partial W}{\partial R} = 0, \quad \chi = 0, \quad \psi = 0 \tag{131e}$$

where $F_p = (\rho_s/\rho_\ell)f_p/a\omega_s$ is a dimensionless pulling rate, $H_m = (\varepsilon_\ell \sigma T_m^3 a/k_\ell)(T_m/\Delta T)$ is the Biot number of the melt, $\theta_a = (T_a - T_m)/\Delta T$, $Ma_r = (-\partial\gamma/\partial T)\Delta Ta/\alpha_t\mu$ is the Marangoni number, $Re_r = a^2\omega_s/\nu$ is the Reynolds number, $R_c = b/a$ is a dimensionless crucible radius, and $D = d/a$ is a dimensionless crucible depth. Usually, the dimensionless pulling rate is assumed to be zero. Then the flow is mainly characterized by seven dimensionless numbers: the Prandtl number Pr, the Reynolds number Re_r, the Grashof number $Gr_r = g\beta \Delta Ta^3/\nu^2$, the Marangoni number Ma_r, the rotation ratio s, the Biot number H_m, the dimensionless crucible radius Rc, and the dimensionless crucible depth D.

Practically, the effect of the flow on the mass transfer is stressed. The effects of the flow on the heat transfer have also been recognized to be important. It is desirable to know the temperature field as well as the flow in the crucible. Various works directed to that end are summarized.

1. Pure Forced Convection

The flow in the crucible is characterized by four dimensionless numbers: the Reynolds number Re_r, the rotation ratio s, the dimensionless crucible radius R_c, and the dimensionless crucible depth D.

a. CRYSTAL ROTATION ALONE

Forced convection caused by crystal rotation alone [44] is shown in Fig. 8. The fluid flows radially outward along the melt free surface and axially downward along the wall of the crucible. For low Reynolds numbers, the flow is confined to the upper part of the crucible and an Ekman layer is formed under the crystal. For high Reynolds numbers, forced convection becomes large in size and occupies the entire crucible. At the same time, a central column, which is sometimes called a Taylor column, appears under the crystal where the fluid flows upward directly from the crucible bottom to the crystal. Another Ekman layer is formed at the crucible bottom as well as at the interface. The critical Reynolds number at which the Ekman layer is formed at the crucible bottom depends on the crucible size. It has been obtained experimentally as a function of the crystal radius [39]. The upward central flow is enhanced by slow crucible rotation and is suppressed by thermal free convection. The critical Reynolds number has been found numerically [44] to be $\sim$700 for R_c = 2.5 and D = 5. The flow with Re_r = 4036, R_c = 1.53, and D = 0.943 was also obtained [52].

b. CRUCIBLE ROTATION ALONE

Forced convection caused by crucible rotation alone is now considered. The fluid flows peripherically upward along the crucible wall and downward along the axis of symmetry [39]. Under the crystal, a cell called a Taylor column is formed [41]. The fluid in this cell does not rotate but flows downward. The fluid outside of the cell rotates at the crucible rotation rate as if it were a solid body. Ekman layers are again formed at the interface and at the crucible bottom. The flow pattern is quite similar to thermal free convection except for the rotation around the axis.

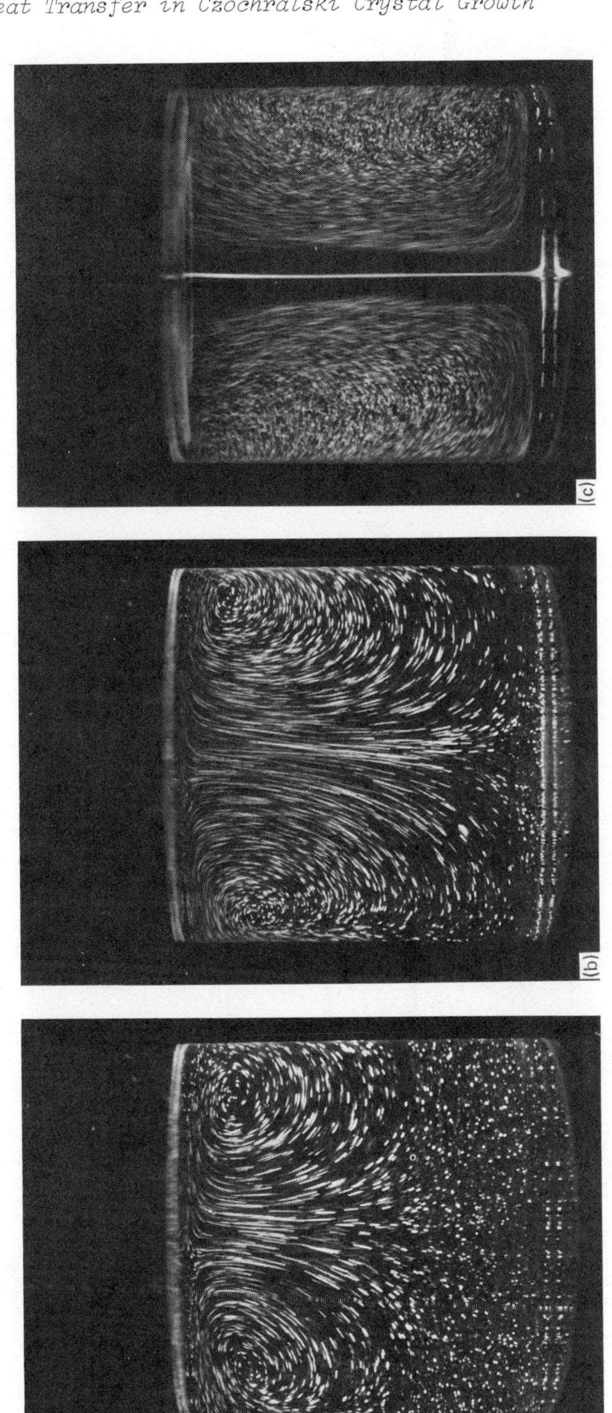

FIG. 8. Simulation of forced convection caused by crystal rotation: (a) $Re_r = a^2 \omega_s / \nu = 51.5$; (b) $Re_r = 196$; (c) $Re_r = 775$.

c. COMBINED CRYSTAL AND CRUCIBLE ROTATION

The crystal rotation controls the flow in the crucible and modifies the shape of the stagnation surface which prevents the fluid in the crucible from mixing completely. It also changes the mode of the heat transfer. Schematic flow patterns [41] are shown in Fig. 9.

When the crystal and the crucible rotate in the same sense, there appears a stagnation surface which forms the Taylor column under the crystal. In the Taylor column, the flow is essentially two-dimensional and independent of the depth. The Taylor column rotates at a rotation rate between those of the crystal and the crucible. In the column, the fluid flows toward the crystal if $1 > s > 0$ and toward the crucible bottom if $s > 1$. Outside the column, the fluid rotates at the crucible rotation rate. When $s \approx 0$, the stagnation surface actually vanishes.

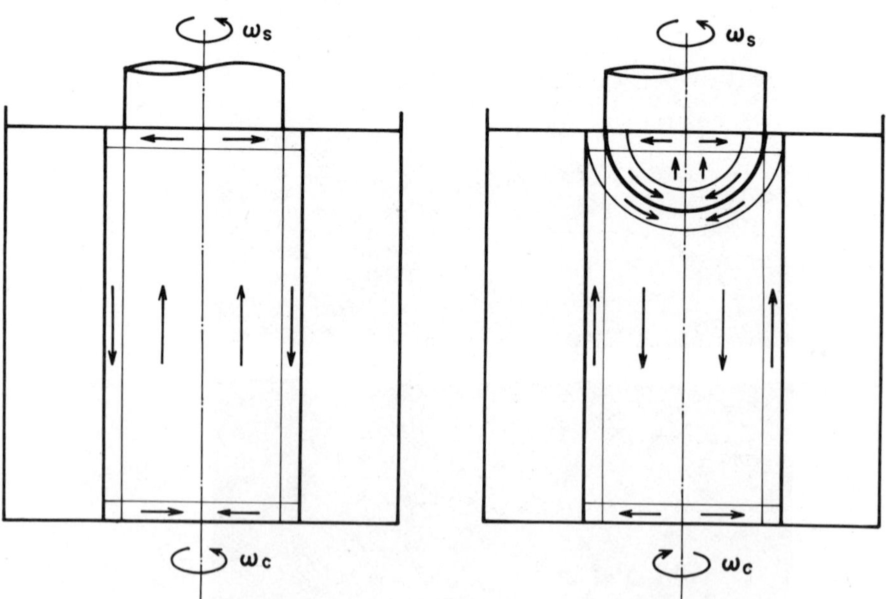

FIG. 9. Schematic flow patterns for combined crystal and crucible rotation: (a) isorotation ($1 > s = \omega_c/\omega_s > 0$); if $s > 1$, flow is reversed; (b) counterrotation ($s < 0$). (From Ref. 41; reprinted with permission.)

When the crystal and the crucible rotate in the opposite sense, two stagnation surfaces appear. One forms the Taylor column vertically under the crystal. The other isolates and confines the forced convection caused by crystal rotation under the crystal. If $-1 > s > 0$, a stable flow with well-defined stagnation surfaces is formed in the crucible. If $s < -1$, the flow in the crucible is ill-defined with forced convection caused by crucible rotation tending to be dominant.

The flows have been studied numerically by a time-independent method [47] for low Reynolds numbers ($Re_r = 0-48$, $Re_r' = sRe_r = 0-48$), $R_c = 2.5$, and $D = 5$. The flows have been obtained numerically by a time-dependent method [52] for $Re_r = 2691$, 4036, 13450, $s = -0.612$, -0.680, $R_c = 1.53$, and $D = 0.943$. Unexpectedly the flows are time-dependent and oscillate in time.

2. *Combined Free and Forced Convection*

In this case the flow in the crucible may be characterized by eight dimensionless numbers. The Prandtl number Pr, the Grashof number Gr_r, the Marangoni number Ma_r, and the Biot number H_m are added to the dimensionless numbers characterizing forced convection. In place of the Grashof number Gr_r, a Grashof number $Gr_d = g\beta \, \Delta T d^3/\nu^2 = Gr_r D^3$ is sometimes used for free convection.

a. *PURE FREE CONVECTION*

Free convection is induced by an adverse or destabilizing temperature gradient. For low Grashof numbers, the fluid flows non-axisymmetrically in the upper part of the crucible like a spoke wheel which varies with crystal or crucible rotation rate. The temperature in the spoke region is lower by a few degrees than in the other surface region [43]. For intermediate Grashof numbers, an axial symmetric flow is induced at the vertical crucible walls. The fluid flows radially inward and axially downward near the crystal. For high Grashof numbers, the circulating flow becomes strong. The flow was laminar for $Gr_d < 1.2 \times 10^9$ and was turbulent for $Gr_d > 1.2 \times 10^9$.

The Marangoni flow driven by surface tension gradient is usually small on the earth for Al_2O_3 melt [55].

b. INFLUENCE OF THE CRYSTAL ROTATION ALONE

Flow pattern. The center column formed by crystal rotation collapses if the fluid is radially heated [43]. The flow is divided into three classes [43,48]. For low Reynolds numbers, free convection is dominant and crystal rotation is not effective. For intermediate Reynolds numbers, combined free and forced convection appears. There is a stagnation surface which separates the flow into two parts. The region near the crystal is occupied by forced convection and the remaining region is occupied by free convection. At the dividing surface between free and forced convection, eddies appear [43]. However, they disappear for a melt with high viscosity so that the flow on the melt free surface is circular. For high Reynolds numbers, forced convection is dominant and a single large vortex appears on the melt free surface. The forced convection pattern is confined to the upper part of the crucible with the dividing surface being a few centimeters below the crystal. At the boundary between the fluid and the crucible, eddies are seen for high-viscosity melts [43].

Similar flow patterns have been obtained numerically [45] for Re_r = 0 to 52.4, Pr = 0.31, Gr_d = 7.36 $\times$ 10^3, s = 0, H_m = 0.57, Ma_r = 0, R_c = 2.5, and D = 5. The flow patterns and temperature fields [48] for Re_r = 40, Pr = 0.01, 0.1, 1, Gr_d = 2 $\times$ 10^3 to 2 $\times$ 10^5, s = 0, H_m = 0, Ma_r = 0, R_c = 2.5, and D = 5 have also been obtained and are shown in Fig. 10. For a low Grashof number, free convection is suppressed by forced convection. On the other hand, for a high Grashof number, strong free convection confines the forced convection to a small region near the crystal.

Time-dependent bulk flows have been computed for a GGG melt [54]. The effect of the Grashof number was primarily studied. The parameters chosen were Re_r = 7975, Pr = 0.053, Gr_d = 0-3.74 $\times$ 10^7, s = 0, H_m = 2.884, Ma_r = 0, R_c = 1.5, and D = 0.9234. All of the solutions evolved to steady states. As the Grashof number increases, the flow caused by the buoyant force forms near the corner of the crucible bottom and suppresses the forced convection. With a further increase of the Grashof number, the forced convection is separated into two parts.

One is a cylindrical region between the crystal and the crucible. The
other is directly under and near the periphery of the crystal. Free
convection occupies the annular region, with width being equal to the
melt height. This shows that the flow pattern is markedly dependent
on the aspect ratio of the crucible.

After all, for a given Reynolds number, the flow is mainly forced
convection for low Grashof numbers and mainly free convection for high
Grashof numbers. There is a critical Grashof number Gr_d^c at which the
flow changes from forced convection to free convection. It depends
on the Reynolds number and crucible size relative to the crystal. If
the Reynolds number is varied, there is similarly a critical Reynolds
number Re_r^c which depends on the Grashof number. Flow in the crucible
depends mainly on the Reynolds number and the Grashof number. The
importance of the free convection as compared to forced convection is
effectively determined by the ratio Gr_d/Re_r^2.

Carruthers [56] analyzed the driving force of the flow and estimated the critical Reynolds number. Free convection was assumed to
be driven by the horizontal temperature inhomogeneity with a driving
force proportional to a Grashof number $Gr_h = g\beta\,\Delta Tb^3/\nu^2$. On the other
hand, forced convection is induced by crystal rotation and its driv-
ing force was assumed to be proportional to the Reynolds number
$Re_r = a^2\omega_s/\nu$. By eliminating the dependence of the kinematic viscosity from Gr_h and Re_r, one obtains

$$Re_r^c = Gr_h^{1/2} \tag{132}$$

Temperature field. As shown in Fig. 10, the temperature in the
crucible depends markedly on the Prandtl number (which represents the
importance of the heat transfer by convection relative to that by
conduction). For low-Prandtl-number melts, the heat is transferred
mainly by conduction. The temperature field is influenced little by
the flow and is mainly determined by the boundary conditions. On
the other hand, for high-Prandtl-number melts, the temperature is
influenced markedly by the flow. In this case two distinct types of
temperature field can be obtained, depending on the mode of the fluid

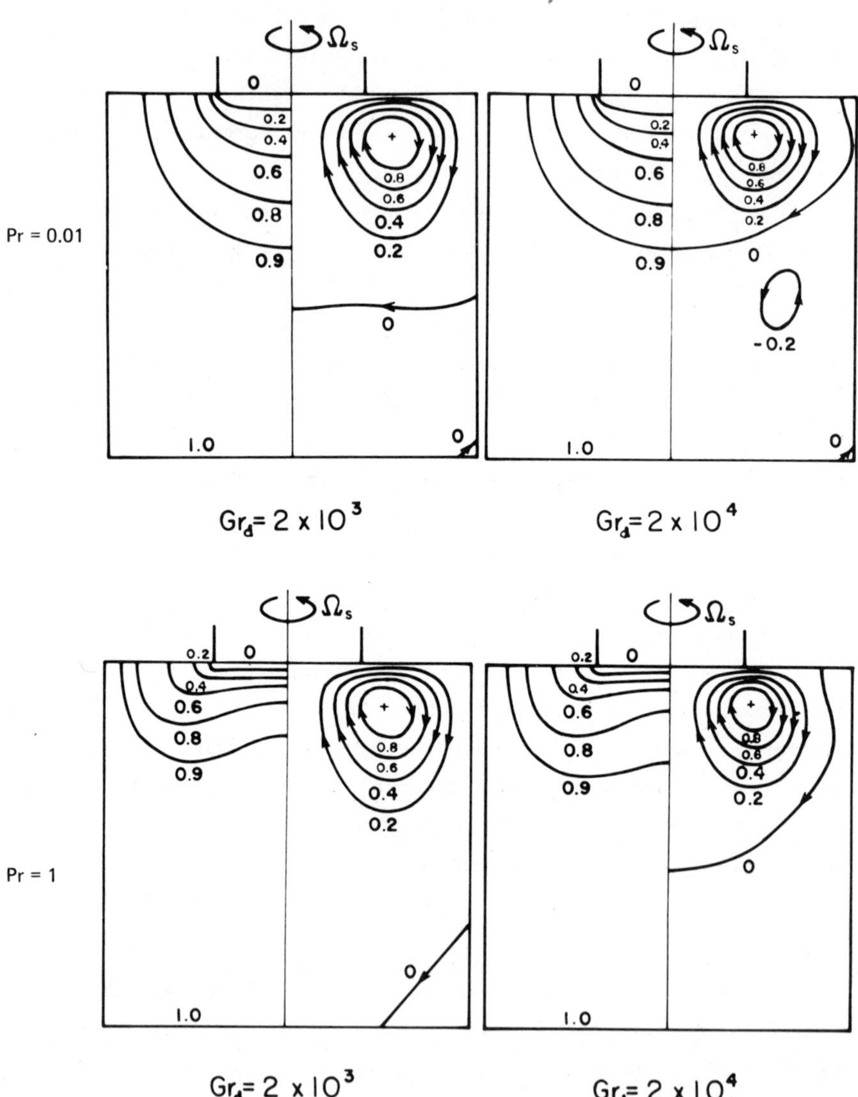

FIG. 10. Flow pattern (right side) and temperature distribution (left side) in a crucible for various Prandtl and Grashof numbers.

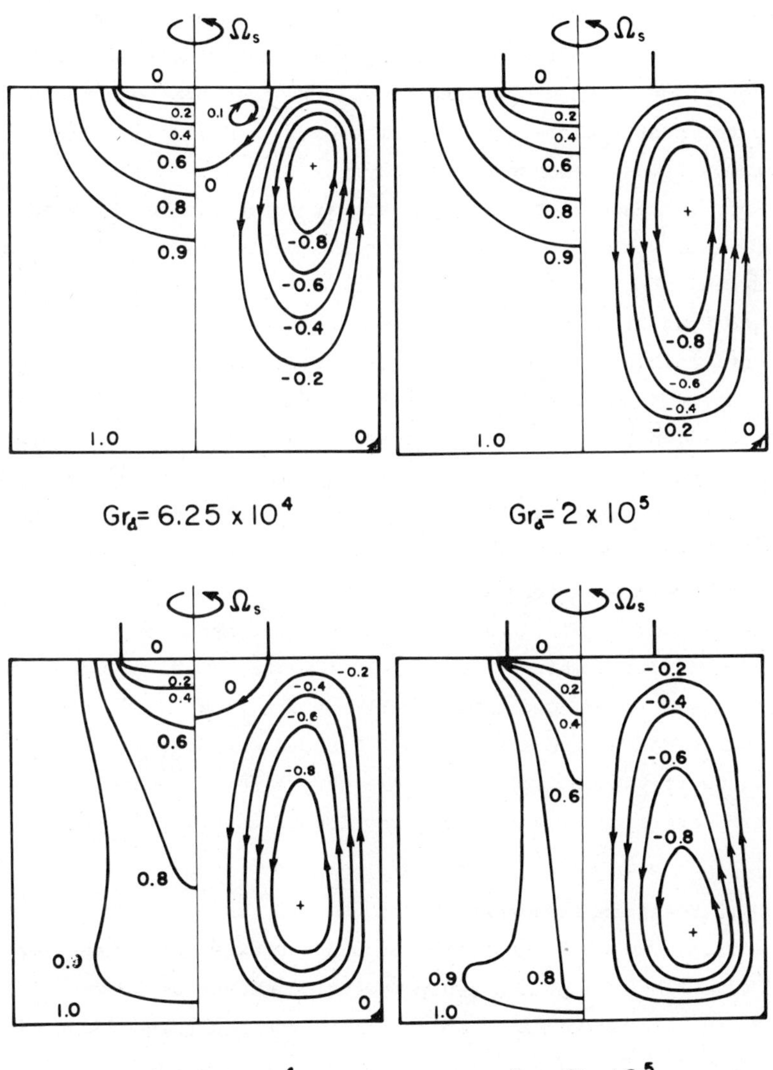

$$Gr_d = 6.25 \times 10^4 \qquad Gr_d = 2 \times 10^5$$

$$Gr_d = 6.25 \times 10^4 \qquad Gr_d = 2 \times 10^5$$

convection [48]. Forced convection pushes the isotherms upward to the crystal and enhances the heat transfer at the interface, while free convection draws them downward toward the crucible bottom and diminishes the heat transfer at the interface.

The temperature field under the crystal varies as the crystal rotation rate increases. Three types of variation can be obtained [42]. For melts with a low Prandtl number, the temperature field is almost independent of the crystal rotation rate. For melts with an intermediate Prandtl number, the temperature varies gradually and free convection is gradually replaced by forced convection as the crystal rotation rate increases. For high-Prandtl-number melts, the temperature suddenly increases by a fixed amount at a critical Reynolds number corresponding to the time when the forced convection due to crystal rotation touches the crucible walls. At the same time, the temperature of the crucible wall suddenly decreases because the heat is more effectively transferred to the fluid from the crucible wall.

The temperature field in molten GGG has also been obtained [53]. Although the Prandtl number is relatively low (Pr = 0.0353), the temperature field is influenced considerably by the flow. Under the crystal, the isotherms are stratified by forced convection. In the annular region of the crucible, the temperature reflects the flow and isotherms are stretched in the direction of the flow.

c. INFLUENCE OF THE CRUCIBLE ROTATION ALONE

Crucible rotation induces forced convection which is similar to free convection, except for rotation around the axis. With respect to the meridional flow, there are two principal differences between forced convection and free convection [46]. First, the position of the circulation center of the forced convection is nearer to the axis than for free convection. Second, the position where the axial velocity is a maximum is nearer to the interface for forced convection. If the crucible rotation rate is low, the flow in the crucible is almost unchanged because free convection is dominant. If the crucible rotation rate is high, forced convection overcomes free convection. Therefore, there is a critical crucible rotation rate at which the fluid flow changes from the free convection dominant mode to the forced convection dominant mode.

Forced convection by crucible rotation transfer the relatively cooled fluid near the melt surface toward the crystal and isotherms near the crystal are compressed radially inward and stretched axially downward by forced convection.

d. INFLUENCE OF COMBINED CRYSTAL AND CRUCIBLE ROTATION

First, we describe the situation for crystal and crucible rotation in the same sense. When the crystal and the crucible rotate at the same rotation rate, the flow is a pure rotation around the axis if the free convection is neglected ($Gr_d = 0$). Meridional flow is induced by free convection. As the Reynolds number Re_r increases from zero to $Re_r' = sRe_r$, the meridional flow changes from a crucible dominant mode to a free convection dominant model If the Reynolds number Re_r is above Re_r', forced convection due to combined crystal and crucible rotation becomes dominant over free convection.

When the crystal and the crucible rotate in the opposite sense, their effects are completely opposite and free convection is generally not effective. If the Reynolds number Re_r is below Re_r', forced convection due to the crucible rotation is dominant. The crystal rotation only disturbs the flow near the crystal by canceling the fluid rotation around the axis and thereby effectively enhances the forced convection due to crucible rotation. If the Reynolds number Re_r is above Re_r', forced convection due to combined crystal and crucible rotation appears in the crucible.

Flow patterns have been studied numerically [46] for various combinations of the Reynolds numbers ($Re_r = 0{-}41.9$, $Re_r' = 10.5$). The other parameters were $Pr = 0.31$, $Gr_d = 7.36 \times 10^3$, $H_m = 0.57$, $Ma_r = 0$, $R_c = 2.5$, and $D = 5$. The time-dependent flow has been computed by using realistic parameters for molten Si [52]. The parameters are $Re_r = 4036$, $Pr = 0.08$, $Gr_d = 7.44 \times 10^7$, $S = -0.680$, $H_m = 2.43$, $Ma = 0$, $R_c = 1.53$, and $D = 0.943$. The computation was stopped when the flow evolved to a steady state. However, the flow remains time-dependent. The temperature field exhibits several interesting features.

There is a small region on the melt free surface just beyond the
crystal-melt boundary where the temperature is below the melting
point. This fact may be due to the overestimation of the radiant heat
dissipation from the melt free surface, as discussed in Sec. VII.B.3.
The isotherms show the tendency to bulge upward as the result of a
pronounced upwelling flow. However, the effect of the flow on the
temperature field is slight because of low Prandtl number.

e. *THERMAL FILM THICKNESS*

The influence of operating conditions on heat transfer to the
crystal is now summarized by using the thermal film thickness at
the interface defined by

$$\delta_f = \frac{T_b - T_m}{(\partial T/\partial z)_{z=0}} \tag{133}$$

where T_b is the bulk melt temperature. In the presence of crystal
rotation alone, the thermal film thickness becomes independent of
radial position as the crystal rotation rate increases, as shown in
Fig. 11. In the presence of crucible rotation alone, the thermal
film thickness depends on the radial position as the crucible rotation
rate increases, as shown in Fig. 12. In the presence of combined
crystal and crucible rotation, the thermal film thickness behaves as
shown in Fig. 11. When the crystal and crucible rotate in the same
sense, the thermal film thickness changes similarly to that for $\omega_c = 0$
with crystal rotation rate. When the crystal and crucible rotate in
the opposite sense, the thermal film thickness at the center becomes
a minimum at a crystal rotation rate $\sim(1/3)\omega_c$. This suggests that the
position of the maximum thickness at the center is shifted to the left
by $\sim\omega_c/3$ if the crucible rotation is applied.

For an infinite rotating disk, the boundary layer theory is applied and thermal film thickness becomes equal to the thermal layer
thickness. Therefore, the thermal film thickness is inversely proportional to the square root of the crystal rotation rate and independent
of radial position. If the flow is induced by combined crystal and
crucible rotation, the rotation difference $\Delta\omega = \omega_s - \omega_c$ between the

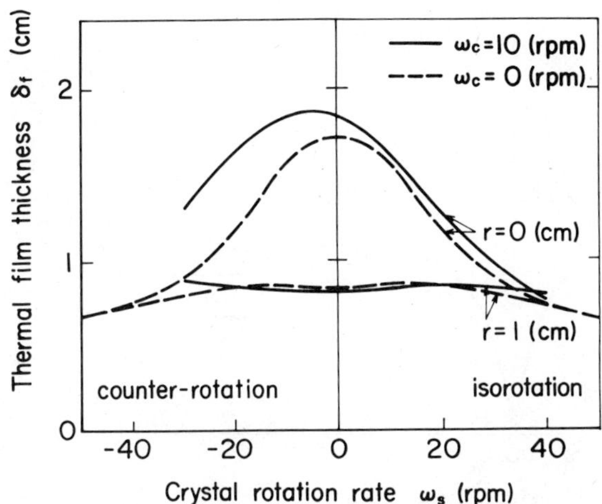

FIG. 11. Thermal film thickness as a function of crystal rotation rate. (Pr = 0.31.)

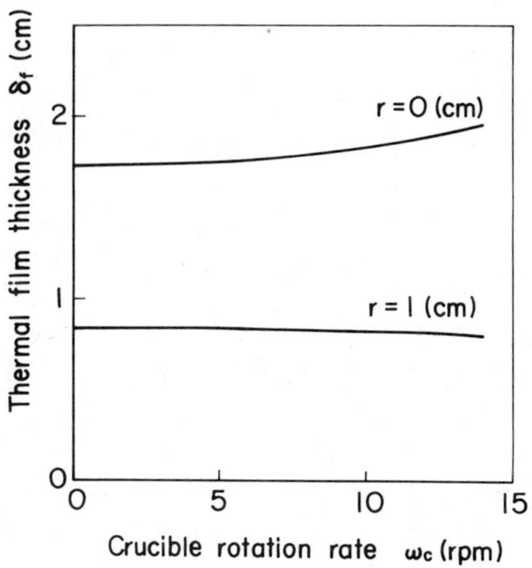

FIG. 12. Thermal film thickness as a function of crucible rotation rat rate. (Pr = 0.31.)

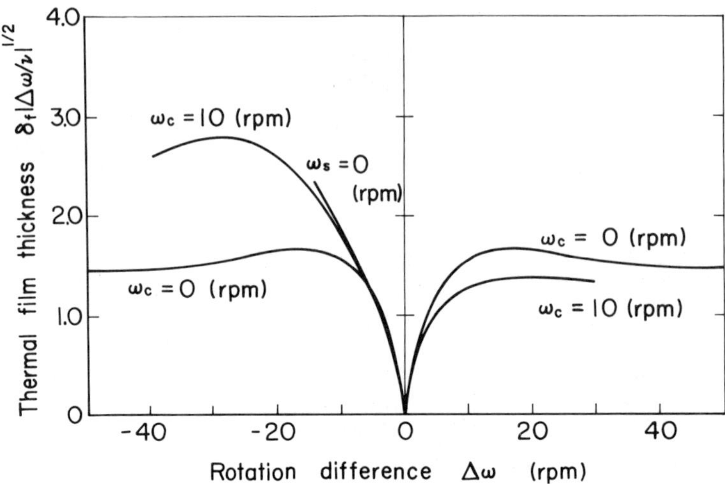

FIG. 13.   Relation between $\delta_f|\Delta\omega/\nu|^{1/2}$ and $\Delta\omega = \omega_s - \omega_c$.

crystal and the crucible should be used in place of the crystal rotation rate ω_s. The quantity $\delta_f|\Delta\omega/\nu|^{1/2}$ is shown in Fig. 13. For high crystal rotation rate, $\delta_f|\Delta\omega/\nu|^{1/2}$ becomes independent of the rotation difference. However, the actual value of the thermal film thickness is about one-third of the thermal layer thickness. This is due to the fact that heat transfer from the crucible wall is efficient.

C. Summary

The effects of the various factors which may be used to control the heat transfer in the melt are summarized.

To decrease the axial temperature in the melt:

1. The crucible temperature is lowered by decreasing the input power.
2. The position of the heat source relative to the crucible is moved upward.
3. A small crucible is used.
4. An afterheater, reflector, or thermal shield is used.
5. The melt is stirred by crucible rotation.
6. The crucible base is cooled.

To increase the axial temperature gradient in the melt:

7. The melt is stirred by crystal rotation unless the crucible base
    is cooled.

VII. INTERFACE SHAPE

Until now we have assumed that the solid-liquid interface shape is
planar. However, in practice a variety of interface shapes are observed. The actual interface shape may be concave, planar, convex,
or some combination of these, ranging from deeply concave to extremely convex. Experimentally, interface shape is usually determined by
observing impurity striations in a grown crystal. Striations are
bands of solute inhomogeneities parallel to the interface because
they are principally induced by fluctuations in convection and in
growth rate [57]. The interface shape can strongly influence the
radial homogeneity of the solute and the dislocation content of the
crystal [58]. It is observed that curved interfaces result in more
dislocations [18,59] or a large residual strain which deteriorates
optical quality [60,61], and even that cracks are sometimes induced
in the crystal leading to catastrophic damage [62,63]. This shows
the importance of controlling the interface shape by controlling the
experimental parameters which influence the interface shape. Practically it is usually desired to have a planar or nearly planar
interface.

One determinant of the deviation of the interface shape from
planarity is the Biot number [12]. Large deviations are typical for
large Biot number. The interface shape becomes more concave (less
convex) with increasing Biot number, unless other conditions are
varied. A concave interface is usually obtained if the melt is completely stirred [12]. A convex interface is obtained if the heat is
transferred mainly radially from the melt to the interface [64] or if
the crystal near the interface absorbs the heat from surroundings such
as the exposed crucible walls [65].

Now, the position of the interface $z = \Gamma(r)$ should be determined by the continuities of temperature and heat flux at the interface:

$$T_s = T_\ell = T_m \tag{134a}$$

$$-k_s\left(\frac{\partial T_s}{\partial n}\right) = -k_\ell\left(\frac{\partial T_\ell}{\partial n}\right) + f_n \rho_s \Delta H \tag{134b}$$

where $\partial/\partial n$ is the partial differential operator normal to the interface from the melt to the crystal and f_n is the normal component of the growth rate to the interface.

A. Crystal Growth from a Completely Stirred Melt

1. A Very Slow Growth Rate

Wilcox et al. [12] considered a model with a constant heat flux q_0 transferred from the melt to the crystal through a disk at the hot end. Latent heat was neglected. This model is the same as the previously described model in Sec.V.B. except for the boundary condition at the disk, which is now

$$-\frac{\partial \theta}{\partial Z} = Q_0 \quad \text{at } Z = 0$$

$$\theta = 1 \quad \text{at } R = 1, Z = 0 \tag{135}$$

where $Q_0 = q_0 a/k_s(T_m - T_a)$ is the dimensionless heat flux. The solution is given by

$$\frac{\theta}{Q_0} = \sum_{n=1}^{\infty} \frac{1}{\alpha_n(H^2 + \alpha_n^2)} \frac{J_0(\alpha_n R)}{J_0(\alpha_n)} \exp(-\alpha_n Z) \tag{136}$$

$$Q_0 = \frac{1}{\sum_{n=1}^{\infty} 1/[\alpha_n(H^2 + \alpha_n^2)]} \tag{137}$$

with α_n already defined in Sec. V.B. The interface shape is the isotherm of $\theta = 1$ which appears at the corner of the disk ($R = 1, Z = 0$). The interface shape is given approximately by

$$Z = -\frac{1}{\alpha_1} \ln \left[\frac{Q\alpha_1 (H_2^2 + \alpha_1^2) J_0(\alpha_1)}{J_0(\alpha_1 R)} \right] \tag{138}$$

except for the region $R \sim 1$.

Interface shapes computed from Eq. (138) are shown for various Biot numbers in Fig. 14. It is obvious that the Biot number is a good measure of the deviation from planarity. The interface is always concave to the melt, with the degree of concavity increasing as the Biot number increases. For low Biot numbers ($H < 0.1$), the interface shape is nearly planar and the one-dimensional approximation becomes valid. These facts are easily recognized by recalling the definition of the Biot number. The Biot number roughly represents the ratio of the heat dissipated from the crystal surface to that transferred by conduction in the crystal. If the Biot number is large, the interface shape becomes concave to permit heat to be lost more easily to the surroundings. If the Biot number is low, the interface shape becomes planar and heat is mainly conducted through the crystal. The shape of the other isotherms are similar to that of the interface. They are concave if the Biot number is large and nearly planar if the Biot number is small.

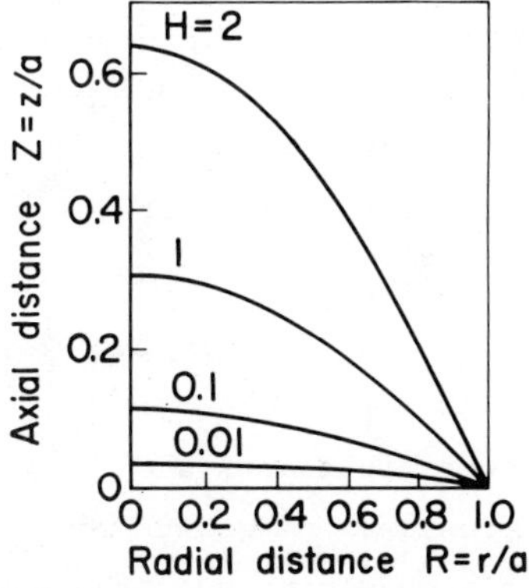

FIG. 14. Interface shapes for various Biot numbers. [From Ref. 66, p. 175, and from W. R. Wilcox and R. L. Duty, *J. Heat Transfer, Trans. ASME, Ser. C, 88;* 47 (1966); reprinted with permission.]

2. *A Finite Growth Rate*

The above model has been refined [12]. A heat balance at the interface was considered with a uniform heat flux from the melt to the interface and latent heat due to the growth at the interface. At the edge of the interface, a radial heat balance gives

$$h_a(T_m - T_a) = \cos[\theta_b \, h_\ell(T_b - T_m) + f\rho_s \, \Delta H \sin \theta_b] \tag{139}$$

where h_ℓ is the heat transfer coefficient from the melt to the interface and θ_b is the angle between the interface and the vertical boundary of the crystal. In Fig. 15, the relation between $S_h = h_\ell(T_b - T_m)/h_a(T_m - T_a)$ and θ_b is shown for various values of $L_h = f\rho_s \, \Delta H/h_a(T_m - T_a)$. For given S_h and L_h, the interface shape will be concave toward the melt if $\theta_b < \pi/2$ and convex if $\theta_b > \pi/2$. This model also always predicts a concave interface under the growth condition $f > 0$ and $T_b > T_m >$ and may be satisfactorily applied to the case that the melt is compJ ɔely stirred by crystal rotation.

A full numerical solution for the interface shape was obtained by solving the differential equations [12] for various Biot numbers and angles of θ_b. Again the Biot number is the prime factor determining the interface shape. The angle θ_b modifies the interface shape slightly, becoming less concave as θ_b increases toward $\pi/2$.

This model is also quite useful for predicting the interface shape when the crystal radius is increasing or decreasing during growth [66]. As schematically shown in Fig. 16, the interface shape is more concave for a constantly increasing diameter than for a constant diameter, and more convex (less concave) for a diameter which increases with time, as experimentally observed [67].

B. Computer Simulation

In actual crystal growth, many different interface shapes are observed. To explain this a new model seemed to be necessary. A growth model [64] is schematically shown in Fig. 17. This model includes the spatial variation of the heat transfer coefficients h_a and h_ℓ, the bulk temperature T_b, and the surrounding temperature T_a. The interface

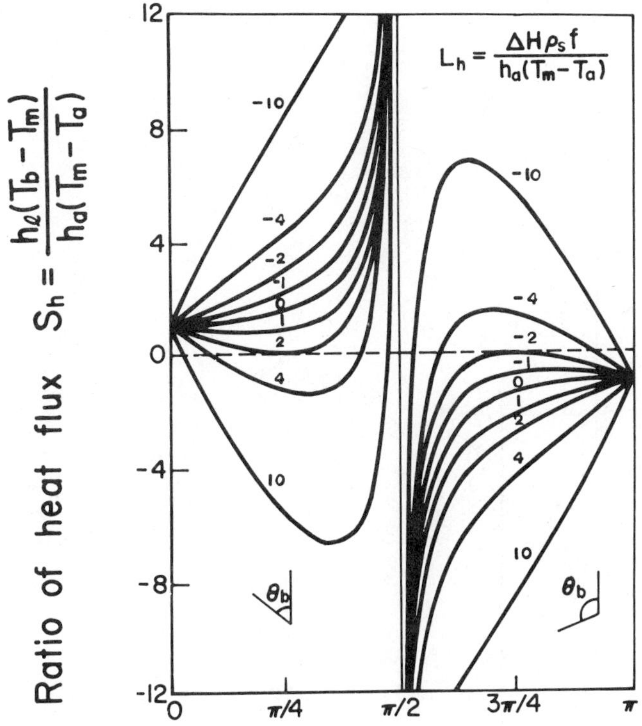

FIG. 15. Relation between S_h and θ_h for various values of L_h (see text). [From Ref. 66, p. 177, and from W. R. Wilcox and R. L. Duty, *J. Heat. Transfer, Trans. ASME, Ser. C, 88*; 46 (1966); reprinted with permission.]

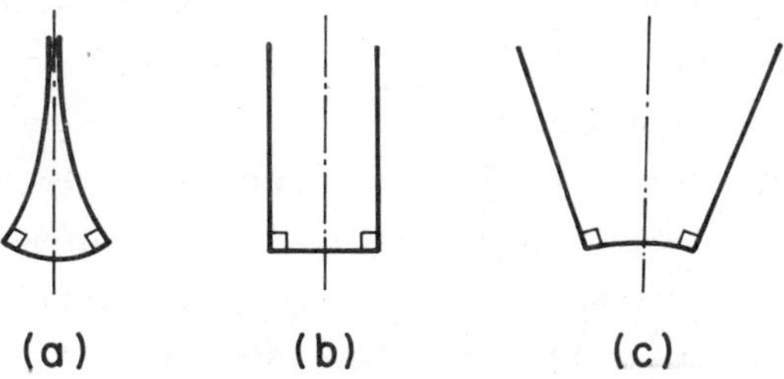

FIG. 16. Interface shape with increasing or decreasing crystal radius. (From Ref. 66, p. 178; reprinted with permission.)

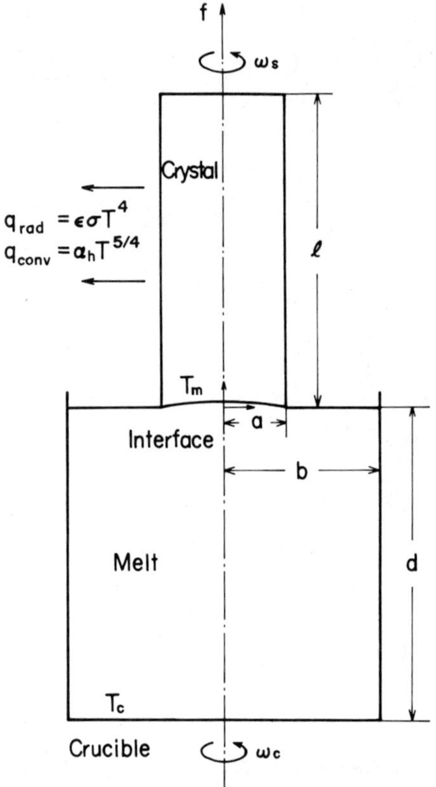

FIG. 17. Crystal growth model.

shape was obtained by solving numerically the differential equations
governing the temperatures both in the solid and in the liquid, using
the method briefly described in Appendix D. For the computations, the
values of many parameters were set beforehand. The parameters are
divided into two groups: material properties and system parameters.
The former include the physical properties such as emissivity, thermal conductivity, melting point, latent heat, specific heat, volume
expansion coefficient, kinematic viscosity, and density. The latter
include the coefficient of heat transfer, crystal radius, crystal
length, crucible radius, melt height, crucible temperature, crystal
rotation rate, and crucible rotation rate. The growth rate is not

explicitly specified because it is uniquely determined by Eq. (134b).
Of course, it may be practical to specify the growth rate. In this
case, either the crucible temperature or the crystal radius is
determined so as to satisfy Eq. (134b). Usually, either method is
not advantageous for saving the computation time.

This method is applied to the growth of Ge [45,46,64]. The heat
dissipation from the crystal surface and from the melt free surface
by radiation and by gaseous convection are given by

$$q_{dis,i} = \varepsilon_i \sigma T_i^4 + \alpha_h T_i^{5/4}, \quad i = s, \ell \tag{140}$$

For simplicity, it is assumed that the radiation is released freely
into the surroundings at zero temperature and that α_h and T_c are constant. The boundary conditions at the crystal surface and at the
melt free surface are given by

$$-k_s \left(\frac{\partial T_s}{\partial z}\right) = q_{dis,s} \quad \text{at the crystal top surface } (0 \leq r \leq a, \; z = \ell) \tag{141a}$$

$$-k_s \left(\frac{\partial T_s}{\partial r}\right) = q_{dis,s} \quad \text{at the crystal side surface } (r = a, \; 0 \leq z \leq \ell) \tag{141b}$$

$$-k_\ell \left(\frac{\partial T_\ell}{\partial z}\right) = q_{dis,\ell} \quad \text{at the melt free surface } (a \leq r \leq b, \; z = 0) \tag{141c}$$

The Biot number is now defined by

$$H = \frac{(\varepsilon_s \sigma T_m^3 + \alpha_h T_m^{1/4})a}{k_s} \tag{142}$$

The effects of the system parameters on the interface shape were
examined because the interface shape is controlled primarily by these
factors. Only one system parameter was varied at a time in order to
isolate its effect on the interface shape. Actually, of course, the
crystal length and melt height vary during a given growth run. The
standard values of the parameters are listed in Table 5. In the present situations, the results are qualitative. The interface shape

TABLE 5

Values of Physical Constants and System Parameters
Used for a Numerical Computation

Physical constant			
Melting point	T_m	937	°C
Latent heat	ΔH	411	$J\ g^{-1}$
Emissivity	ε_s	0.2	
	ε_ℓ	0.2^a	
Thermal conductivity	k_s	0.24	$W\ cm^{-1}\ K^{-1}$
	k_ℓ	0.71	$W\ cm^{-1}\ K^{-1}$
Specific heat	$c_{p\ell}$	0.4^a	$J\ g^{-1}\ K^{-1}$
Density	ρ_ℓ	5.5	$g\ cm^{-3}$
Kinematic viscosity	ν	0.1^a	$cm^2\ sec^{-1}$
Thermal expansion coefficient	β	0.0001^a	K^{-1}
Derivative of surface tension	$-(\partial\gamma/\partial T)$	0^a	$N\ cm^{-1}\ K^{-1}$

System parameter (standard value)			
Coefficient of heat transfer	α_h	0	$W\ cm^{-2}\ K^{-5/4}$
Crystal radius	a	1	cm
Crystal length	ℓ	5	cm
Crucible radius	b	2.5	cm
Melt depth	d	5	cm
Crucible temperature	T_c	943	°C
Crystal rotation rate	ω_s	0	rpm
Crucible rotation rate	ω_c	0	rpm

[a] Assumed.

itself is not conclusive because it can vary by changing the thermal
conditions. The trend of the variation of the interface shape is
meaningful. Then the words "more concave (less concave)" also mean
"less convex (more convex)" in this section. The results are also
useful for the other materials if the dimensionless numbers are used.
The dimensionless numbers are also listed in Table 6.

TABLE 6

Investigated Ranges of Dimensionless Numbers

Dimensionless number	Definition	Range
Biot number	$H = \dfrac{\varepsilon_s \sigma T_m^3 a + \alpha_h T_m^{1/4} a}{k_s}$	0.00418–0.0231
	$H_m = \dfrac{\varepsilon_\ell \sigma T_m^3 a}{k_\ell} \dfrac{T_m}{T_c - T_m}$	0.214–0.571
Prandtl number	$Pr = \dfrac{\nu}{k_\ell / \rho_\ell c_{p\ell}}$	0.31
Reynolds number	$Re_r = \dfrac{a^2 \omega_s}{\nu}$	0–52.4
	$Re_r' = \dfrac{a^2 \omega_c}{\nu}$	0–15.7
Grashof number	$Gr_d = \dfrac{g\beta (T_c - T_m) d^3}{\nu^2}$	58.9–19600
Maragoni number	$Ma_r = \dfrac{(-\partial\gamma/\partial T)(T_c - T_m)^2}{(k_\ell / \rho_\ell c_{p\ell})\mu}$	0
Crucible radius	$R_c = \dfrac{b}{a}$	1.5–3.6
Melt depth	$D = \dfrac{d}{a}$	1–10

1. *No Fluid Flow in the Melt*

This model [64] is useful for the melts with low Prandtl numbers because the heat is mainly transferred by conduction. The effects of
ambient gas, crystal radius, crystal length, crucible radius, melt
height, crucible temperature, and growth rate on the interface shape
are examined.

a. *AMBIENT GAS*

The coefficient of heat transfer α_h depends on the species in the surrounding gas and may be estimated by Eq. (12). In this section, α_h is varied as a parameter. Obviously, $\alpha_h = 0$ W cm^{-2} K$^{-5/4}$ corresponds to growth in a vacuum. As α_h increases, the interface shape becomes more concave, as shown in Fig. 18. The growth rate for fixed crucible temperature also increases with α_h. These results are qualitatively explained by the variation of the Biot number.

Unfortunately, experimental results are not explicitly reported.

b. *CRYSTAL RADIUS*

Interface shapes for various crystal radii are shown in Fig. 19. The degree of concavity increases with crystal radius up to 1.5 cm (presently, 60% of the crucible radius) and rapidly decreases as the crystal radius approaches the crucible radius. For a large crystal radius near the crucible radius, the interface shape would be convex, though it is not explicitly shown, because heat is transferred mainly radially from the melt to the interface. The corresponding growth rate monotonically decreases with the crystal radius.

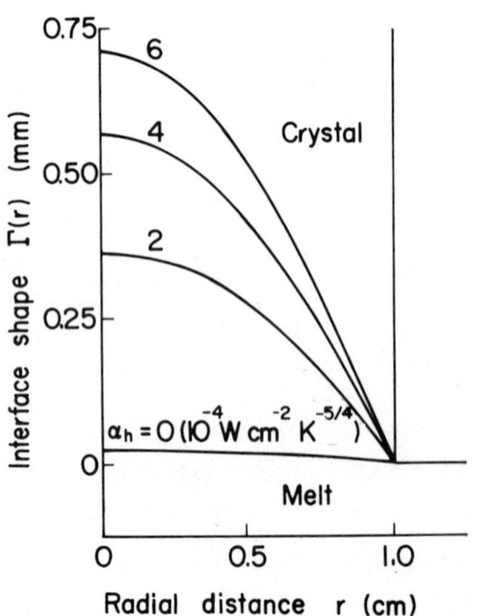

FIG. 18. Interface shapes for various coefficients of heat transfer.

For a large crystal radius, the Biot number is no longer a good measure for predicting the deviation of the interface shape from planarity. This is due to the finiteness of the crucible radius. If the crystal radius is small compared to the crucible radius, the Biot number remains the primary determinant of the interface shape.

Experimentally, as crystal radius is increased, the interface shape becomes more concave (less convex) and sometimes even changes from convex to concave [67-69].

c. CRYSTAL LENGTH

As the crystal grows, crystal length increases and at the same time the melt height decreases. As discussed in Sec. VII.B.1.e, the effect of the melt height on the interface shape is slight unless the melt height is small.

The interface shapes for various crystal lengths are shown in Fig. 20. As the crystal length increases, the degree of concavity of the interface shape decreases. The interface shape approaches that for infinite length as crystal length increases. The critical length for which the interface shape may be regarded as almost independent

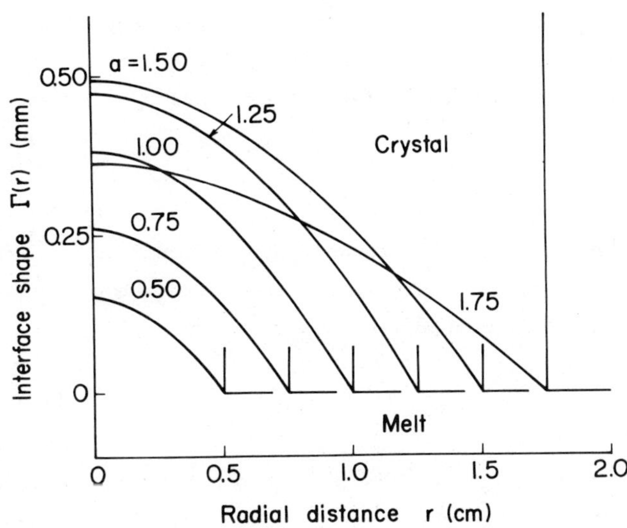

FIG. 19. Interface shapes for various crystal radii (in cm).

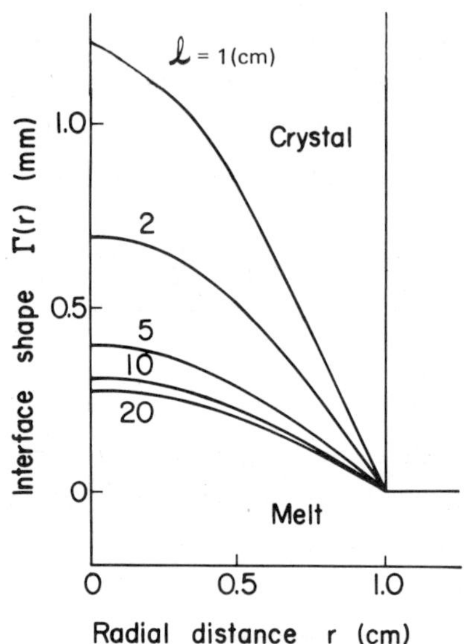

FIG. 20. Interface shapes for various crystal lengths.

of crystal length has already been shown in Sec. V. The growth rate
for a given crucible temperature also increases with crystal length.
With increasing crystal length, the heat dissipated from the crystal
surface increases and so the growth rate increases for a constant
crystal radius. If a fixed growth rate is desirable, the crucible
temperature can be varied with crystal length, as mentioned in Sec.
VII.B.1.g.

These results may also be understood by considering a Biot number
which includes the effect of the crystal length. For a finite crystal,
the heat transferred by conduction is roughly proportional to the temperature difference between the interface and the crystal top surface
$\Delta T_e = T_m - T_e$. Then the Biot number may be defined by

$$H_f = H \left(\frac{T_m}{\Delta T_e} \right) \tag{143}$$

Thus a short crystal is more concave than a long one because the Biot
number is larger for the shorter crystal.

Experimentally, the interface shape typically changes from concave to convex during growth [70]. This is mainly due to the effects

of the increasingly exposed crucible wall. The effect of the increasing crystal length on the interface shape is slight.

d. CRUCIBLE RADIUS

This section is closely related to the previous Sec. VII.B.1.b.
Really, the effect of the crystal radius on the interface shape is
now considered from another point of view. The interface shape is
shown as a function of crucible radius in Fig. 21. As the crucible
radius decreases, the interface shape becomes less concave, while it
is convex for a small crucible radius. This shows that a convex interface shape is obtained naturally if the heat is transferred mainly
radially from the melt to the interface. If the melt is heated from
the sides, the interface shape will be convex. The growth rate for
fixed crucible temperature decreases as the crucible radius decreases.
This is due to the increases of the heat flux from the melt to the
interface.

Unfortunately, experimental results are not explicitly reported.

e. MELT HEIGHT

As the crystal grows, melt height decreases unless raw material
is continuously supplied. As the melt height decreases, the interface

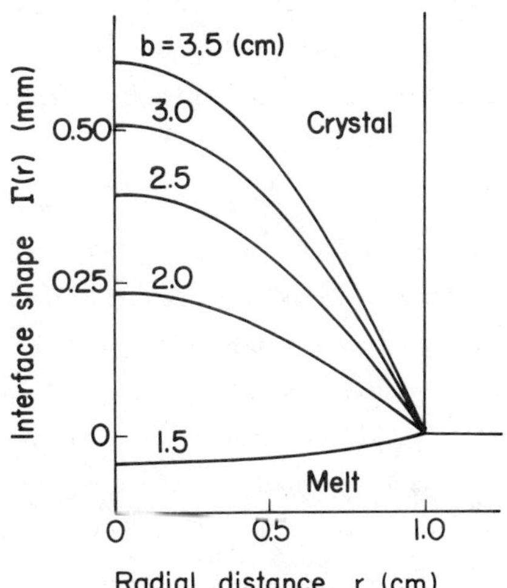

FIG. 21. Interface shapes for various crucible radii.

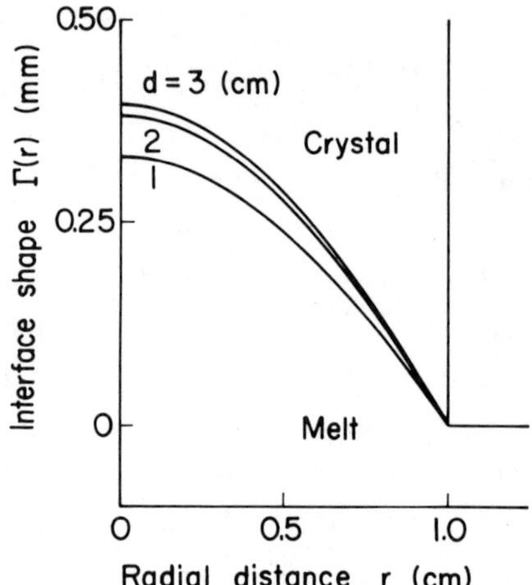

FIG. 22.   Interface shapes for various melt heights.

shape becomes increasingly less concave, reflecting the shape of
the crucible bottom, as shown in Fig. 22.   For a deeper melt, the interface shape becomes more concave and approaches that for an infinite
melt height.   The growth rate corresponding to fixed crucible temperature decreases with decreasing melt height because the heat transferred
from the crucible bottom increases.

Experimentally, the interface shape changes from concave to convex as the melt height drops [18,68].   This result is also due to the
effect of an increasingly exposed crucible wall.   The effect of the
melt height on the interface shape is relatively small.

f.   CRUCIBLE TEMPERATURE

The interface shape is shown as a function of crucible temperature in Fig. 23.   The interface shape changes from concave to convex
as the crucible temperature increases.   A convex interface shape is
again obtained for a high crucible temperature and may be explained
by the increasingly radial heat transfer from the melt to the interface.   The deviations of the interface shape from planarity for various
heat transfer coefficients are shown in Fig. 24.   The interface shape
becomes less concave with increasing crucible temperature.   An
increased crucible temperature corresponds to a diminished growth
rate, as shown in Fig. 25.

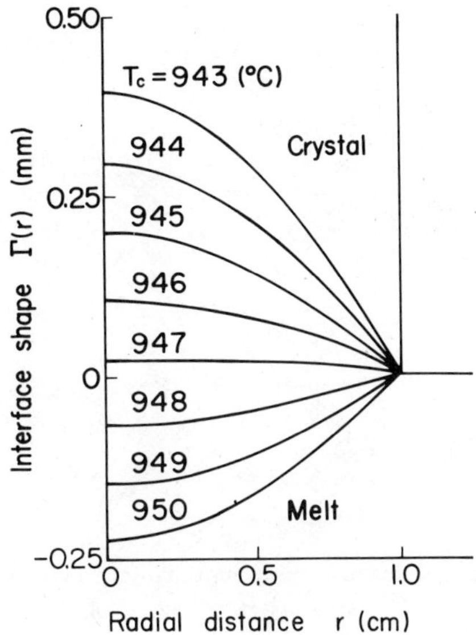

FIG. 23.   Interface shapes for various crucible temperatures.

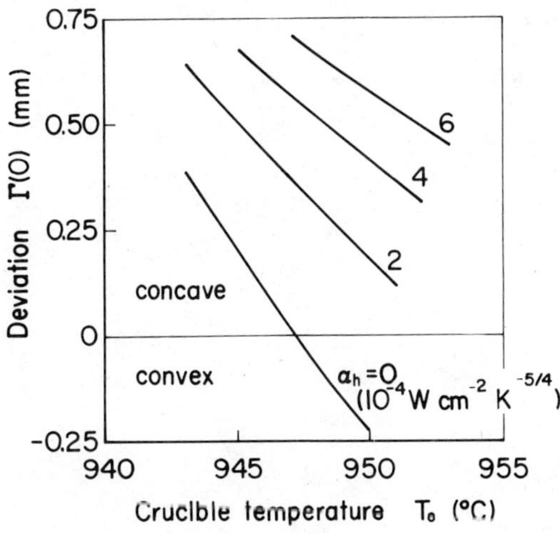

FIG. 24.   Deviations of the interface shapes from planarity for various coefficients of heat transfer.

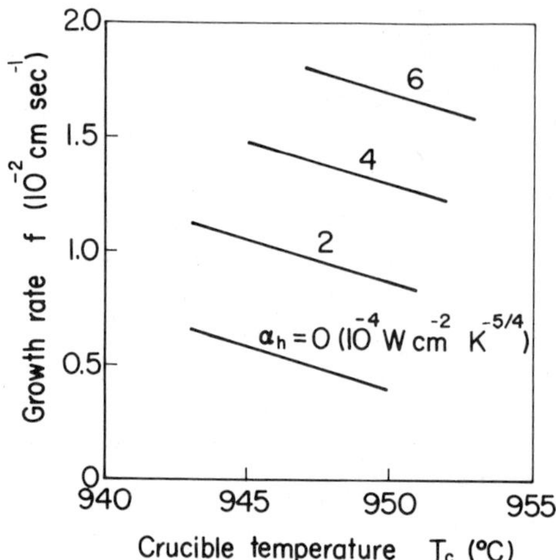

FIG. 25. Growth rates as functions of crucible temperature for various coefficients of heat transfer.

g. GROWTH RATE

The permissible growth rate is controllable by the crucible temperature, as shown in Sec. VII.B.1.f. Experimentally, it is well known that the growth rate considerably influences the interface shape. Mil'vidskii and Golovin [68] empirically obtained a useful relation between the growth rate and the interface shape, as illustrated in

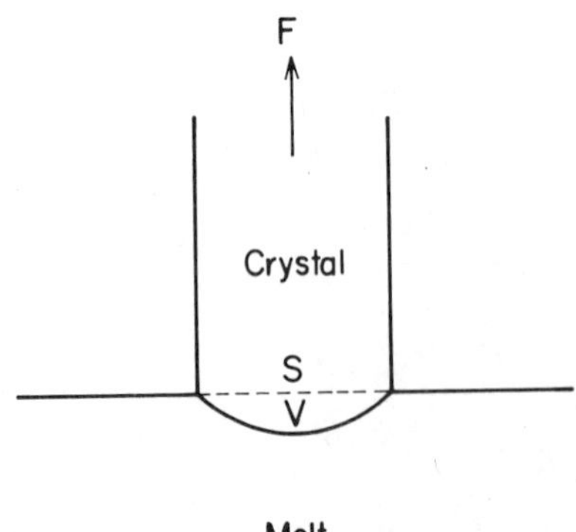

FIG. 26. Crystal growth model by Mil'vidskii and Golovin. (From Ref. 68; reprinted with permission.)

Fig. 26. A crystal with convex interface is imagined to be growing at the growth rate f. They found the growth rate f_0 at which the crystal grows with a planar interface,

$$f_0 = f + c_0 \left(\frac{V}{S}\right) \tag{144}$$

where c_0 is a constant ($c_0 = 1 \text{ sec}^{-1}$ if the growth rate is given in mm sec^{-1}), V is the capped volume of the interface, and S is the cross-sectional area of the crystal. If the quantity V/S is the height of a cylinder having volume V, then Eq. (144) means that the convexity of the interface vanishes if the growth rate is increased by V/S per unit time. Of course, the sign of the volume V is negative if the interface shape is concave.

The relation between growth rate f and V/S for various coefficients of heat transfer is shown in Fig. 27. Each profile is nearly linear except for large growth rates, and so can be expressed by

$$f = -c_1 \left(\frac{V}{S}\right) + c_2 \tag{145}$$

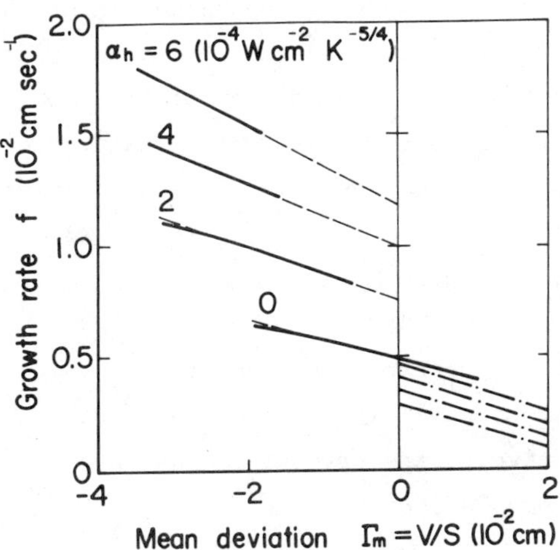

FIG. 27. Relation between the growth rate and the mean deviation and the results obtained by Mil'Vidskii and Golovin (dash-dotted lines). (From Ref. 68; reprinted with permission.)

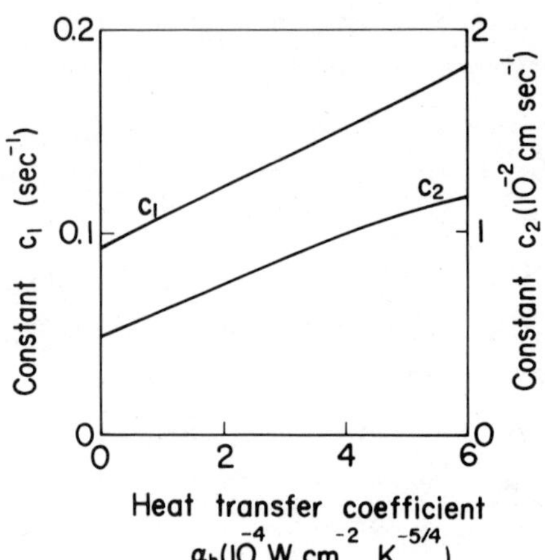

FIG. 28. Parameters c_1 and c_2 as functions of heat transfer coefficient.

where c_1 and c_2 are constants determined from the straight lines in Fig. 26 and are shown in Fig. 28. If $c_1 = 0.1$ sec^{-1}, Eqs. (144) and (145) coincide with one another. However, the constant c_1 usually depends on the heat dissipation from the crystal surface and is a measure of the stability of the interface shape to the variation of the growth rate. If c_1 is large, the interface is stable to changes in growth rate. In other words, it is difficult to control the interface shape by a change of the growth rate. On the other hand, the constant c_2 represents the growth rate of the crystal with a planar interface shape and in fact is the same as f_0 in Eq. (144). If a planar interface shape is desired, a growth rate equal to c_2 (or f_0) must be used.

Experimentally, the interface shape becomes less convex and sometimes changes from convex to concave with increasing growth rate [37,67-69]. The deviation relative to the crystal radius is almost independent of growth rate for metals [68].

2. *Fluid Flow in the Melt*

The previous results obtained by neglecting the fluid flow will be altered considerably if the fluid flow in the melt is taken into

account. Especially the effects of the crystal radius, crucible radius, melt height, and crucible temperature on the interface shape
are markedly influenced by the flow in the melt if the Prandtl number
is not small. However, qualitative features are not altered drastical-
ly. In this section, we consider the situation in which flow in the
melt strongly influences the interface shape. The examined factors
are the crystal rotation and the crucible rotation. For computational
convenience, an artificial melt is used [45,46]. A high viscosity of
0.1 cm^2 sec^{-1} is used in place of that of Ge (0.0013 cm^2 sec^{-1}) because the computational method described in Appendix D fails to give
any solution for low kinematic viscosity. Therefore, the results are
only qualitative, but will be effective if dimensionless numbers
are used. One supposed difference between the present solution and
that for a real melt is that the present solution gives a diminished flow because it is caused by a lower Reynolds number and Grashof
number, corresponding to a smaller crystal radius, slower crystal
rotation rate, and lower temperature difference between the crystal
and the crucible. The present interface shape is supposedly more
influenced by the flow because the Prandtl number is larger. The
various fluid flows [45,46] are shown in Fig. 29.

a. CRYSTAL ROTATION ALONE

The interface shapes for various crystal rotation rates are shown
in Fig. 30. As the crystal rotation rate increases, the interface
shape becomes more concave and the corresponding growth rate decreases.
This is because the crystal rotation enhances the radial uniformity
of the temperature field near the crystal and at the same time increases the heat transfer from the melt to the interface. For large
crystal rotation rates, the interface shape approaches that obtained
for a heat transfer coefficient which is uniform across the interface,
as already obtained in Sec. VII.A.

Experimentally, the interface shape changes from convex to concave with increasing crystal rotation rate [37,63,71]. The crystal
rotation rate at which the interface shape becomes planar depends on
crystal size, crucible size, and crucible temperature [60,61]. When

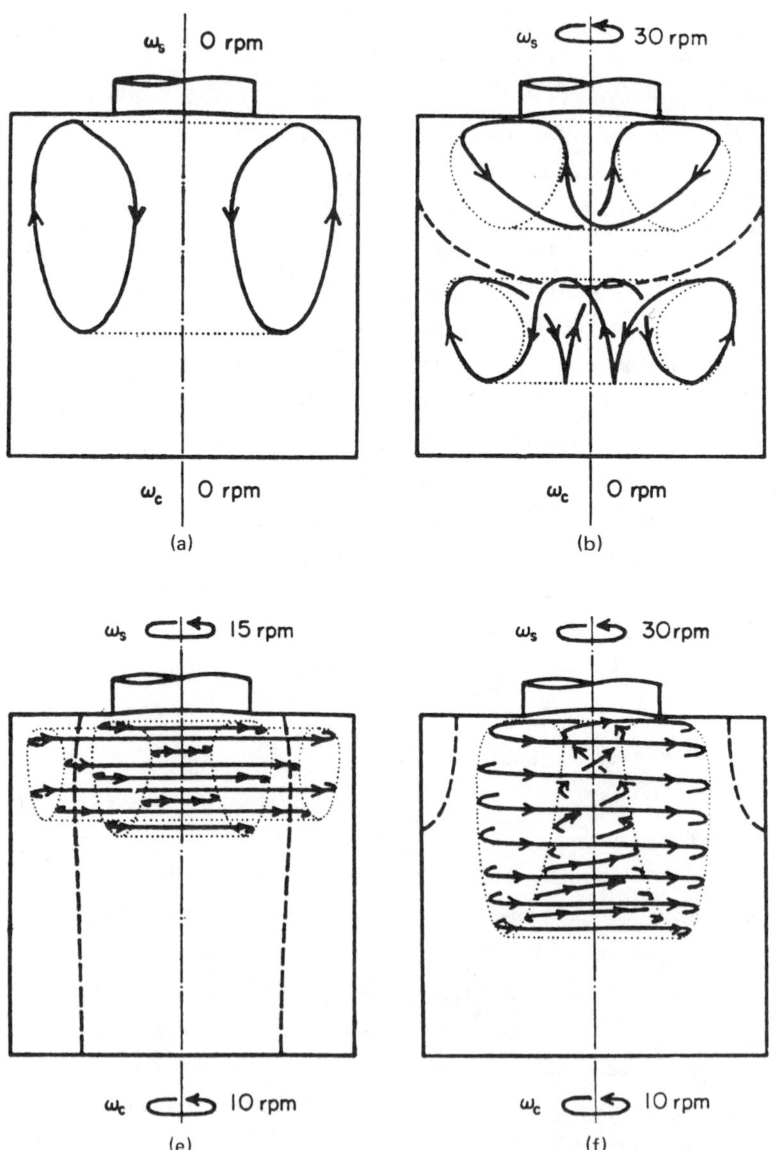

FIG. 29. Flow patterns obtained by computer simulation.

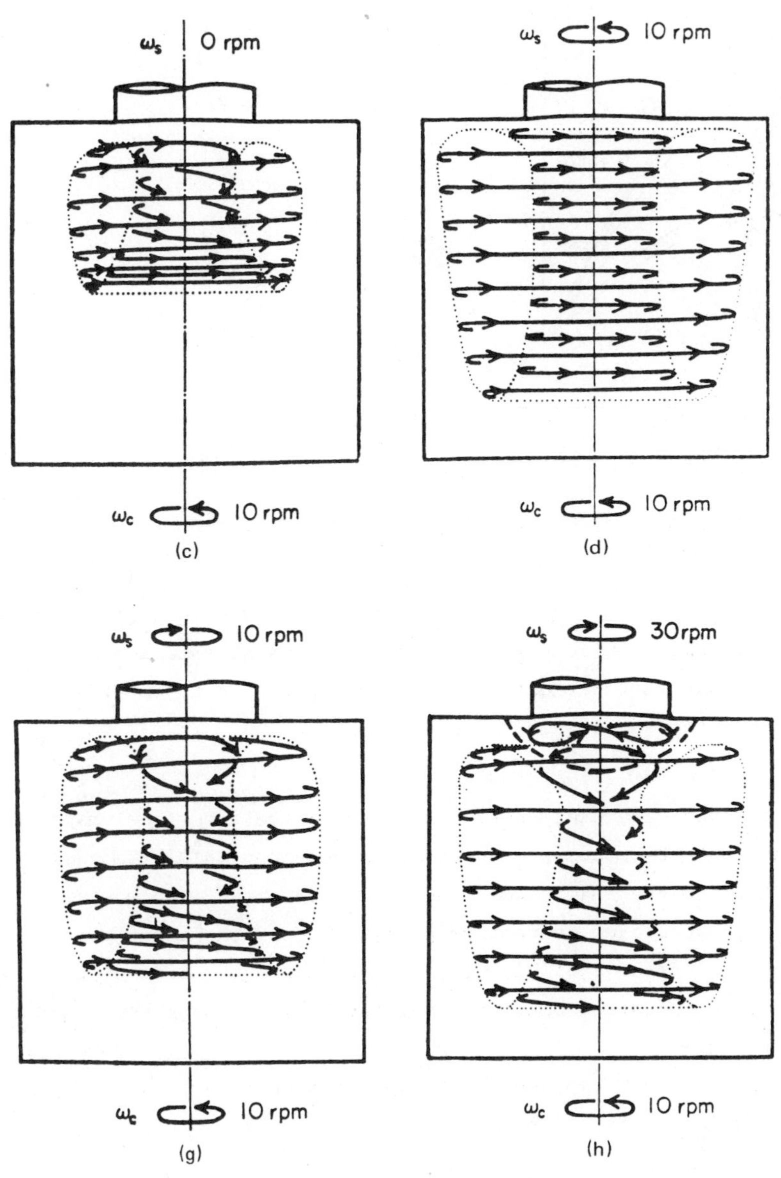
ω_s 0 rpm
ω_c 10 rpm
(c)
ω_s 10 rpm
ω_c 10 rpm
(d)
ω_s 10 rpm
ω_c 10 rpm
(g)
ω_s 30 rpm
ω_c 10 rpm
(h)

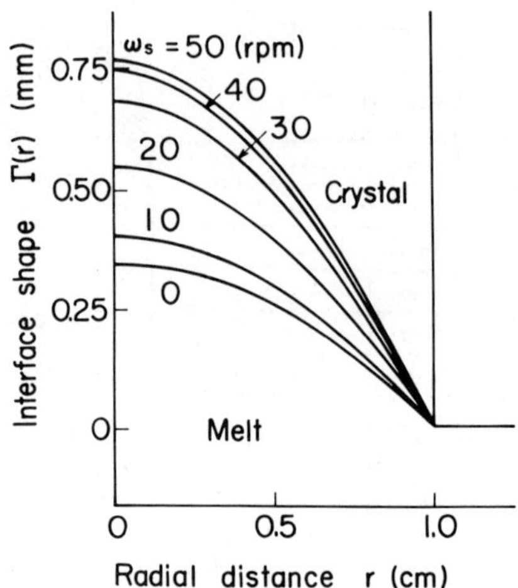

FIG. 30. Interface shapes for various crystal rotation rates.

the crystal grows from a shallow melt in a large crucible, the interface shape is slightly convex even for high crystal rotation rates
[2]. This shows that the effect of crystal rotation is considerably
diminished in the shallow melt.

b. *CRUCIBLE ROTATION ALONE*

As the crucible rotation rate increases, the interface shape becomes less concave because the heat is transferred mainly radially
from the melt to the interface. The interface shape is almost unchanged as long as free convection predominates. The corresponding
growth rate for fixed crucible temperature is almost constant for low
crucible rotation rates, but increases slightly for high crucible
rotation rates.

Experimentally, the interface shape becomes more convex with increasing crucible rotation rate [67].

c. *COMBINED CRYSTAL AND CRUCIBLE ROTATION*

In this section, a crucible rotation rate of 10 rpm (1.047 rad
$\sec^{-1}$) is taken as a typical example of a rate that is sometimes used.
The deviation of the interface shape from planarity is shown as a
function of crystal rotation rate in Fig. 31. When the crystal and

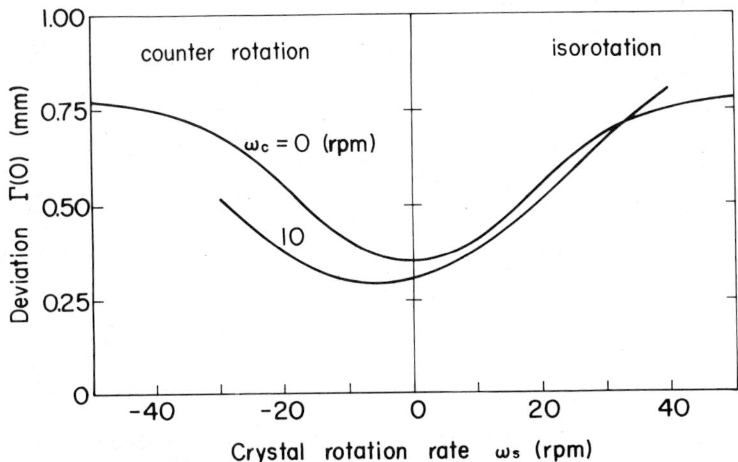

FIG. 31. Deviations of the interface shape from planarity for $\omega_c = 0$ and $\omega_c = 10$ rpm.

the crucible rotate in the same sense, the interface shape becomes more concave as the crystal rotation rate increases. The corresponding growth rate for fixed crucible temperature decreases with the crystal rotation rate. When the crystal and the crucible rotate in the opposite sense, the interface shape becomes less concave initially and then becomes more concave with increasing crystal rotation rate. The least concave interface shape is obtained at a crystal rotation rate of $\sim$3 rpm. The corresponding growth is nearly constant for the crystal rotation rate.

3. Refinement of the Model

In the previous crystal growth model, the heat dissipation from the crystal and melt free surface was overestimated, though a number of the essential features of the crystal growth were qualitatively explained. However, refinement of the model is indispensable to predict the crystal growth processes more exactly. The previous model made it clear that the prime factor controlling the heat transfer is the heat dissipation. Now the heat flow is reexamined and the heat dissipation is evaluated more precisely.

Radiative Heat Dissipation

In the presence of various surfaces with different emissivities and temperatures, the radiative heat dissipation is roughly obtained from Eq. (16). For brevity, the radiative heat exchanges among the crystal surface, the melt free surface, and the exposed crucible wall are considered as shown in Fig. 32. The heat flux radiated from the crystal side surface is given by

$$q_{rad,s}(z) = \varepsilon_s \sigma [\varepsilon_\ell f_{s\ell}(z)(T^4 - \bar{T}_\ell^4) + \varepsilon_c f_{sc}(z)(T^4 - T_c^4)$$

$$+ f_{sa}(z)(T^4 - T_a^4)] \tag{146}$$

where $f_{s\ell}$, f_{sc}, and f_{sa} are the geometric factors viewed from a point on the crystal side surface to the melt free surface, to the exposed crucible wall, and to the other surroundings, respectively; $\bar{T}_\ell$ is the average temperature of the melt free surface and defined in Appendix B; ε_c is the emissivity of the crucible; and T_a is the surrounding temperature. The view factors are shown for various exposed crucible wall height in Fig. 33(a). The view factor $f_{s\ell}$ quickly decreases as the axial position becomes far from the melt free surface. The view factor f_{sc} increases with the wall height h_w and is a maximum at the position $\sim h_w/2$. On the other hand, the view factor f_{sa} decreases with wall height and is given by

$$f_{sa} = 1 - f_{s\ell} - f_{sc} \tag{147}$$

The radiative heat dissipation from the crystal top surface is given by

$$q_{rad,e}(r) = \varepsilon_s \sigma (T^4 - T_a^4) \tag{148}$$

where the crystal top is assumed to be out of the crucible ($f_{sa} = 1$).

The radiative heat dissipation from the melt free surface is given by

$$q_{rad,\ell}(r) = \varepsilon_\ell \sigma [\varepsilon_s f_{\ell s}(r)(T^4 - \bar{T}_s^4) + \varepsilon_c f_{\ell c}(r)(T^4 - T_c^4)$$

$$+ f_{\ell a}(r)(T^4 - T_a^4)] \tag{149}$$

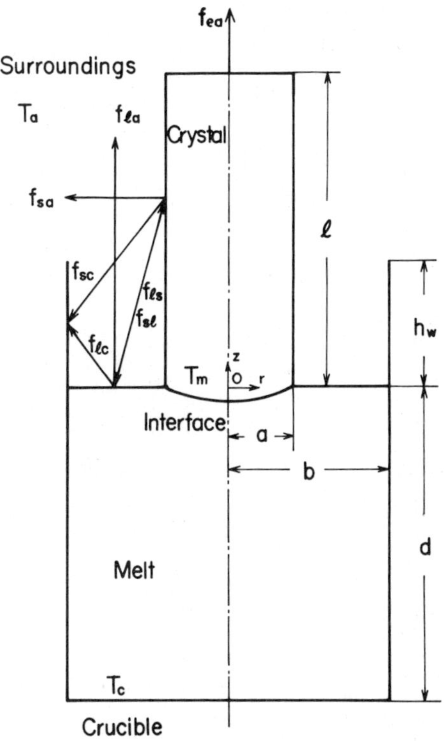

FIG. 32. Schematic illustration of radiative heat exchange during
crystal growth.

where $f_{\ell s}$, $f_{\ell c}$, and $f_{\ell a}$ are the geometric factors viewed from a point
on the melt free surface to the crystal side surface, to the exposed
crucible wall, and to the other surroundings, respectively; and $\bar{T}_s$
is the average temperature of the crystal side surface. These are defined in more detail in Appendix B. The view factors are shown for
various exposed wall heights in Fig. 33(b). The view factor $f_{\ell s}$ monotonically decreases with exposed wall height and is a maximum at the
foot of the wall. On the other hand, the view factor $f_{\ell a}$ decreases
with exposed wall height and is given by

$$f_{\ell a} = 1 - f_{\ell s} - f_{\ell c} \tag{150}$$

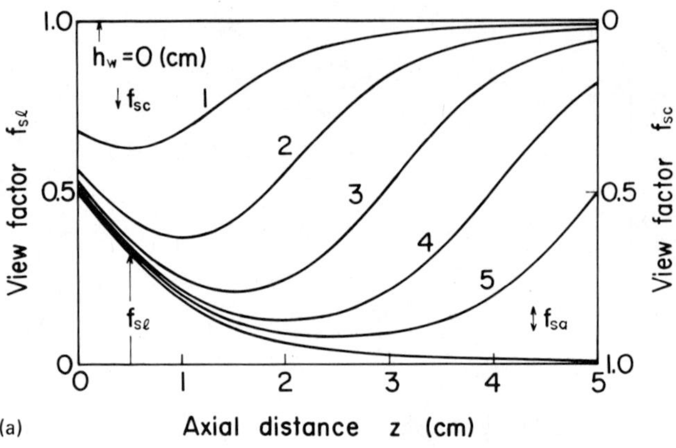

(a) **Axial distance   z   (cm)**

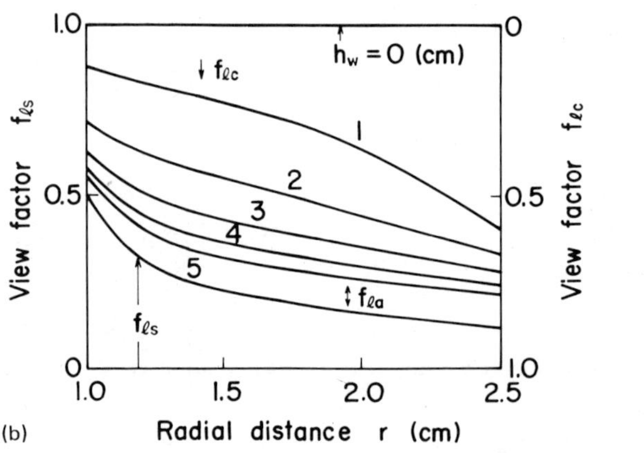

(b) **Radial distance   r   (cm)**

FIG. 33. Geometric view factors: (a) from the crystal side surface
to the melt free surface $f_{s\ell}$, to the exposed crucible wall f_{sc}, and
to the surroundings f_{sa}, for various values of exposed crucible wall
height; (b) from the melt free surface to the crystal side surface
$f_{\ell s}$, to the exposed crucible wall $f_{\ell c}$, and to the surroundings $f_{\ell a}$,
for various values of exposed crucible wall height.

Convective Heat Dissipation

Usually the crystal is grown in an atmosphere of a gas such as
hydrogen, helium, argon, etc. This ambient gas may transfer the heat
from the crystal or the melt in two ways. In one, the heat is forcibly
transferred from the surface by blowing the gas jet onto the crystal.

In the other, heat is transferred by thermal free convection of the ambient gas. The main role of the ambient gas is to prevent the materials in the furnace from oxidizing or evaporating. Heat dissipation by gaseous convection is not now neglected with respect to the radiative heat dissipation.

The heat dissipation from the crystal surface and the melt free surface by convection is given by Eq. (11):

$$q_{conv,i} = h(T - T_a), \quad i = s, e, \ell \tag{151}$$

where h and T_a may now depend on position. For forced convection, h depends on the direction, speed, and flow pattern of the gas flow. Usually such a gas flow is not applied during crystal growth. For free convection, h is usually expressed by

$$h = \alpha_h (T - T_a)^{1/n} \tag{152}$$

where α_h is the coefficient of heat transfer, $n = 4$ for laminar flow, and $n = 3$ for turbulent flow. Strictly speaking, the coefficient should be obtained by solving the entire gas flow problem in the furnace. From the standpoint of a practical computation, α_h is simply a parameter. The validity of Eq. (152) should be judged experimentally.

The values assumed for numerical computation were the same as those used in the previous Sec. VII.B.1. The parameters newly added are the height of the exposed crucible wall h_w, the surrounding temperature T_a, and the emissivity of the crucible ε_c. For brevity, the fluid flow in the melt is neglected here because the Prandtl number is low. The effect of the wall height on the interface shape is mainly studied. The value used for the wall height will be **explicitly** shown whenever appropriate. It is assumed that $T_a = 300$, $\varepsilon_c = 1$, and $n = 4$.

a. IN A VACUUM

To clearly show the difference between the previous model and the refined model, the net heat fluxes radiated from the crystal side surface and the melt free surface are shown in Fig. 34. For convenience of comparison, radiation efficiency is defined by

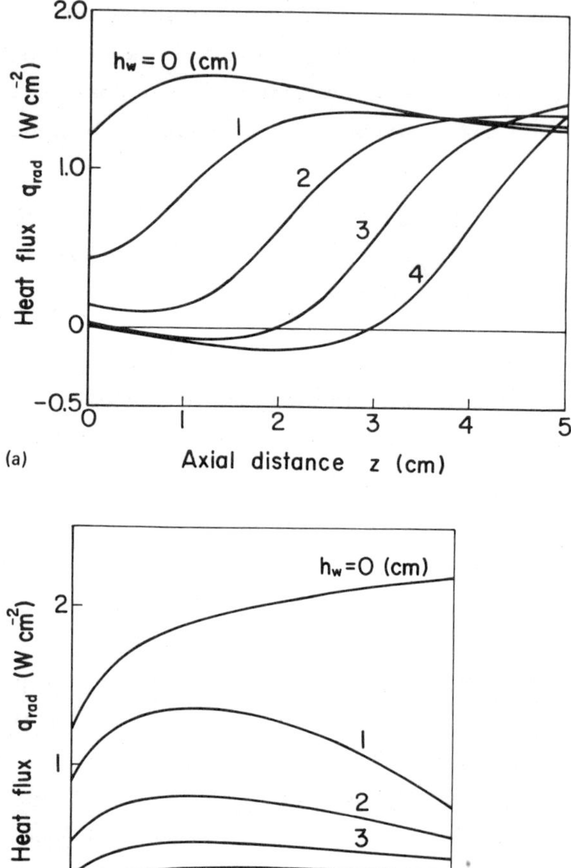

FIG. 34. Heat fluxes radiated (a) from the crystal side surface, and
(b) from the melt free surface, for various values of exposed crucible
wall height. (T_a = 300 K.)

$$\eta_i = \frac{q_{rad,i}}{\varepsilon_i \sigma T^4}, \quad i = s, \ell \tag{153}$$

In the previous model, the radiation efficiency was always unity for

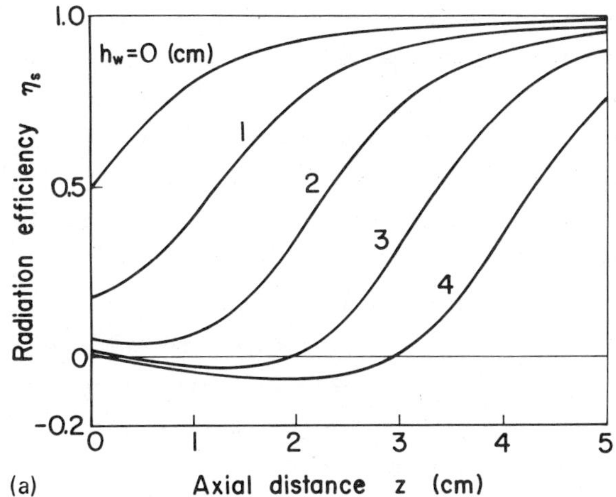

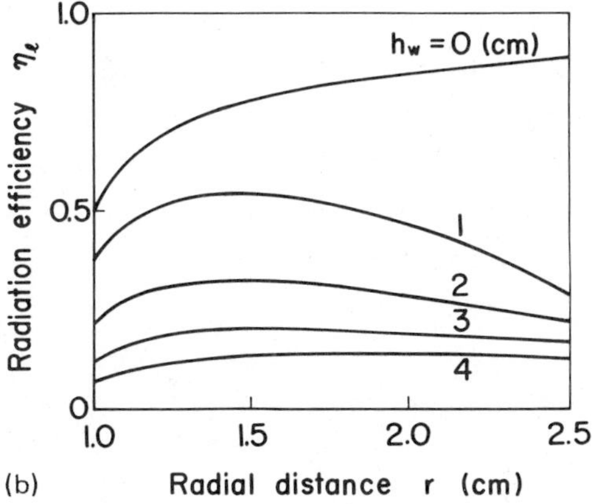

FIG. 35.   Radiation efficiencies of (a) the crucible side surface, and (b) the melt free surface.

the crystal and melt free surfaces and was taken to be independent of position. As shown in Fig. 35, the radiation efficiency η_s of the crystal surface at a point within the crucible ($z < h_w$) becomes considerably smaller. If the wall height is higher than ∿3 cm, a region of negative efficiency appears within the crucible. There the heat is no longer dissipated but is in fact absorbed from the exposed crucible wall. This is usually observed during the growth of a crystal.

On the other hand, the efficiency η_ℓ of the melt free surface becomes
more uniform over the melt free surface as the wall height increases.

(i) Temperature distribution. As mentioned above, the wall
height has a marked effect on the heat transfer. As the wall height
increases, the temperature slowly varies with axial position. The
wall height behaves as if it were an afterheater. The effect of the
wall height on the temperature distribution explicitly appears in the
axial temperature gradient of the crystal, which is nearly constant
within the crucible, as shown in Fig. 36. Therefore, the temperature
distribution of the crystal within the crucible is almost linear because there the heat dissipation from the crystal surface is almost
neglected. On the other hand, the temperature distribution in the melt
is almost unchanged except near the melt free surface.

(ii) Interface shape. Figure 37 shows interface shapes for various values of exposed crucible wall height. The interface shape
is slightly concave for h_w = 0 cm and becomes convex as the wall height
increases. A convex interface shape is obtained naturally if the
thermal effect of the exposed crucible wall is considered. Because the
the heat flow in the crystal is mainly axial, the interface shape

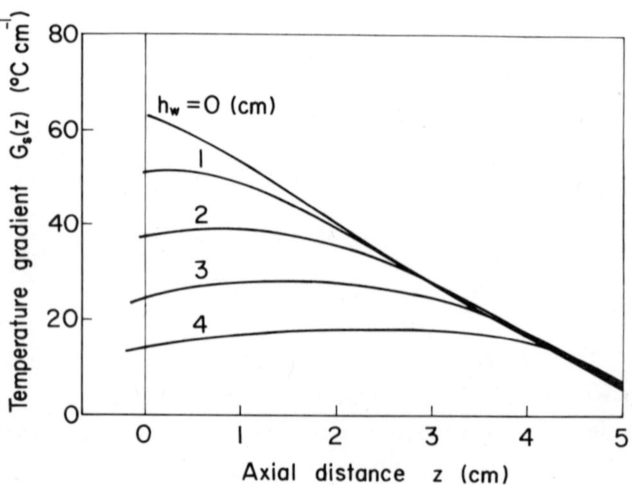

FIG. 36. Axial temperature gradients in the crystal for various values
of exposed crucible wall height. (T_a = 300 K.)

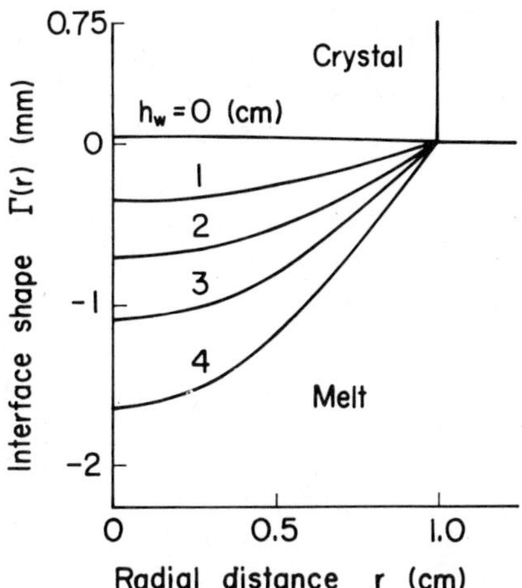

FIG. 37. Interface shapes for various values of exposed crucible wall height. (T_a = 300 K.)

reflects the nonuniformity of the heat flow in the melt and becomes convex. The surrounding temperature has little effect on the interface shape if it is not large.

Growth rates for fixed crucible temperature as functions of exposed crucible wall height are shown in Fig. 38. The growth rate decreases with the wall height. For a high wall, the growth rate becomes negative. In other words, the crystal would actually melt. This is of prime importance for the crystal grower.

b. IN AN AMBIENT GAS

Heat dissipation by gaseous convection has not been extensively studied. In this section, two extreme cases for heat transfer coefficient relative to axial position are examined. An experimental test should be made to see which is the most appropriate model for Czochralski growth.

Constant Heat Transfer Coefficient

The heat flux dissipated from the crystal side surface for α_h = 4 × 10^{-4} W cm^{-2} K$^{-5/4}$ is shown in Fig. 39. The radiative heat dissipation q_{rad} is similar to that in a vacuum and is more influenced

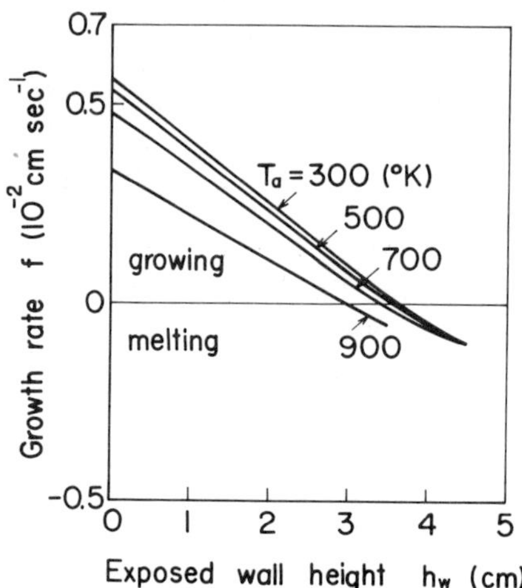

FIG. 38. Growth rates as functions of exposed crucible wall height for various surrounding temperatures.

by the wall height because the crystal is then cooled more by gaseous convection. The portion of the crystal within the crucible absorbs heat mainly by radiation from the exposed crucible wall. The heat dissipated by gaseous convection q_{conv} is a slowly decreasing function of axial position. The total heat flux becomes positive because the heat dissipation by gaseous convection is effective. On the other hand, heat dissipation from the melt free surface is almost unchanged except that a nearly constant heat dissipation by gaseous convection is added.

(i) Temperature distribution. Because the heat is now dissipated from the crystal surface by combined radiation and gaseous convection, the axial temperature gradient becomes steeper and so the temperature rapidly decreases with axial position. The axial temperature gradient in the crystal for α_h = 4 × 1-$^{-4}$ W cm^{-2} K$^{-5/4}$ is shown in Fig. 40. As the wall height increases, the axial temperature gradient at the interface decreases monotonically because of the decrease of the radiative heat dissipation. One of the different characteristics compared to those in a vacuum is that now there is no region with a nearly constant temperature gradient in the crystal.

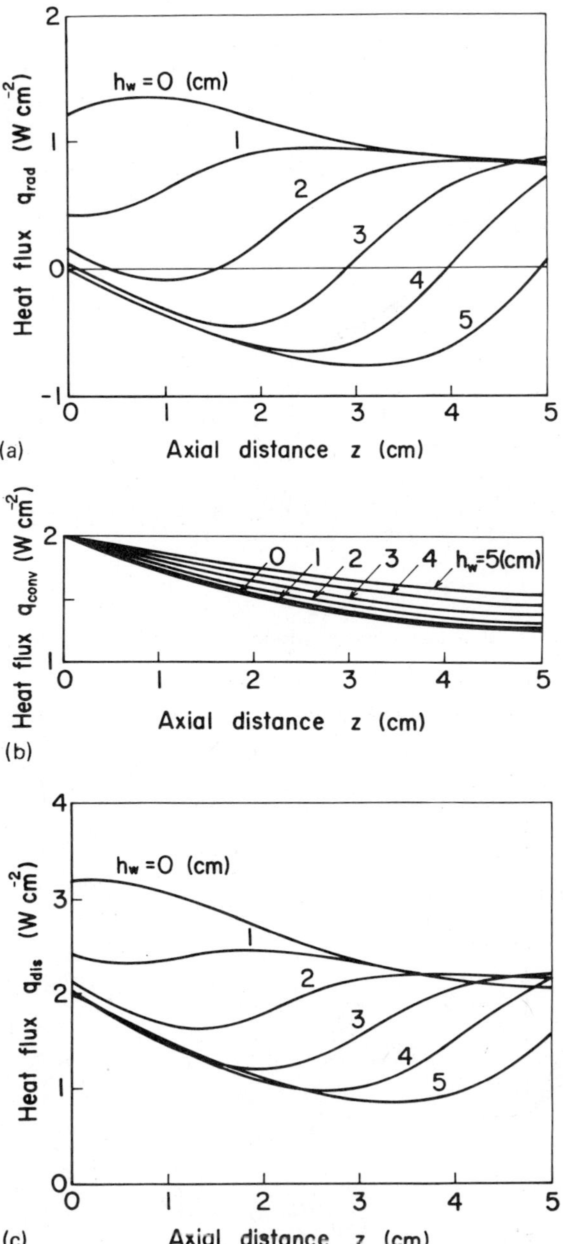

FIG. 39.   Heat fluxes dissipated from the crystal side surface for various values of exposed crucible wall height (a) by radiation, (b) by convection, and (c) by combined radiation and convection. (α_h = 4×10^{-4} W cm^{-2} K$^{-5/4}$ and T_a = 300 K.)

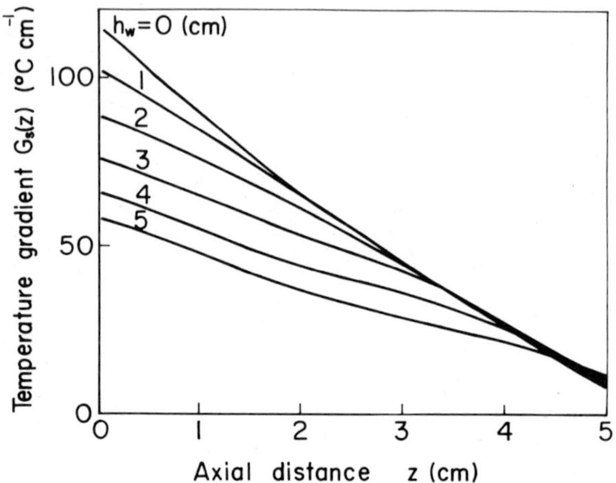

FIG. 40. Axial temperature gradients in the crystal for various values of exposed crucible wall height. ($\alpha_h = 4 \times 10^{-4}$ W cm^{-2} K$^{-5/4}$ and $T_a = 300$ K.)

(ii) Interface shape. The interface shape for $\alpha_h = 4 \times 10^{-4}$ W cm^{-2} K$^{-5/4}$ is shown in Fig. 41. As the wall height increases, initially the interface shape becomes less concave because of the decrease

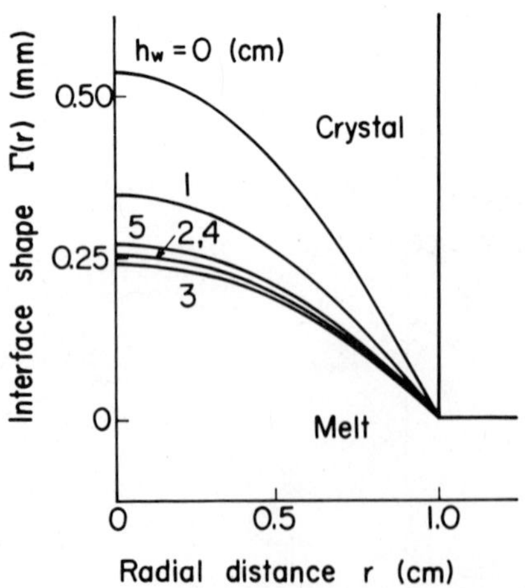

FIG. 41. Interface shapes for various values of exposed crucible wall height. ($\alpha_h = 4 \times 10^{-4}$ W cm^{-2} K$^{-5/4}$ and $T_a = 300$ K.)

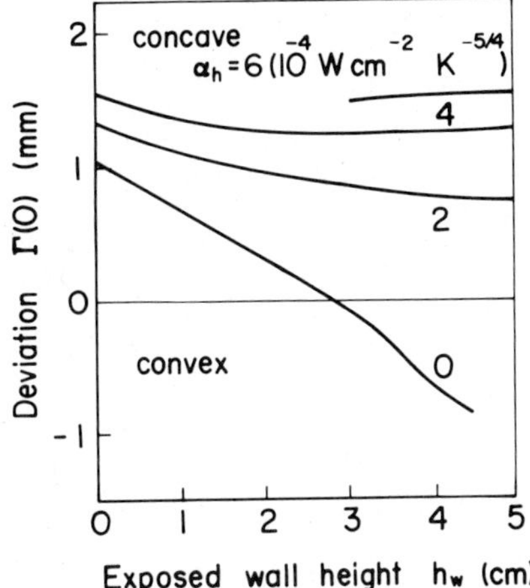

FIG. 42. Deviations of the interface shape from planarity as functions
of exposed crucible wall height for various coefficients of heat transfer. (T_a = 300 K.)

of the radiative heat dissipation. Later it is almost unchanged because then the heat is mainly dissipated by gaseous convection. The
deviations of the interface shape from planarity for various heat
transfer coefficients are shown in Fig. 42. For high heat transfer
coefficients, the variation of the interface shape with wall height
is slight, i.e., the interface shape is stable. The growth rate for
fixed crucible temperature decreases almost linearly with wall height,
as shown in Fig. 43.

Variable Heat Transfer Coefficient

We now consider the situation when a heat transfer coefficient
depends on axial position. As a typical example, a simple function-
ality is used:

At the crystal side surface:

$$\alpha_h(z) = \begin{cases} 0 & z < h_w \\ \alpha_0 & z \geq h_w \end{cases} \qquad (154a)$$

and at the melt free surface:

$$\alpha_h(r) = 0 \quad a \leq r \leq b \qquad (154b)$$

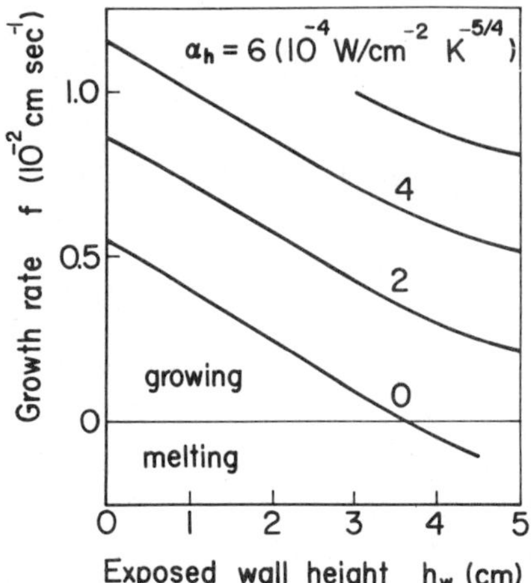

FIG. 43.   Growth rates as functions of exposed crucible wall height for various coefficients of heat transfer.   (T_a = 300 K.)

where α_0 is a fixed value, i.e., heat dissipation by gaseous convection within the crucible is neglected.   This assumption gives results similar to those for a vacuum.   Although the gas temperature is

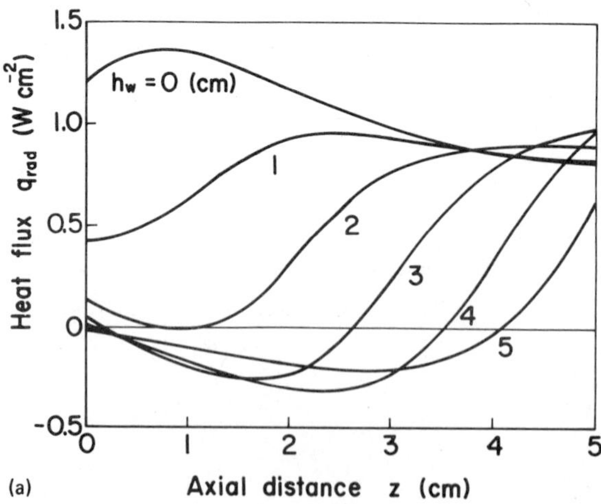

FIG. 44.   Heat fluxes dissipated from the crystal side surface for various values of exposed crucible wall height (a) by radiation, (b) by convection, (c) by combined radiation and convection.   (α_0 = 4 × 10^{-4} W cm^{-2} K$^{-5/4}$ and T_a = 300 K.)   Calculated heat flux from the temperature gradient obtained by Brice is shown by circles.

assumed to be constant, present results are almost equivalent to those for constant heat transfer coefficient and variable gas temperature so that $T_a \simeq T$ within the crucible and T_a is a constant outside the crucible.

The heat flux from the crystal side surface for $\alpha_0 = 4 \times 10^{-4}$ W cm^{-2} K$^{-5/4}$ is shown in Fig. 44. The heat dissipated by radiation is intermediate between that for a vacuum and that for an ambient gas with constant heat transfer coefficient. The heat dissipated by gaseous convection varies stepwise with axial distance. The heat is

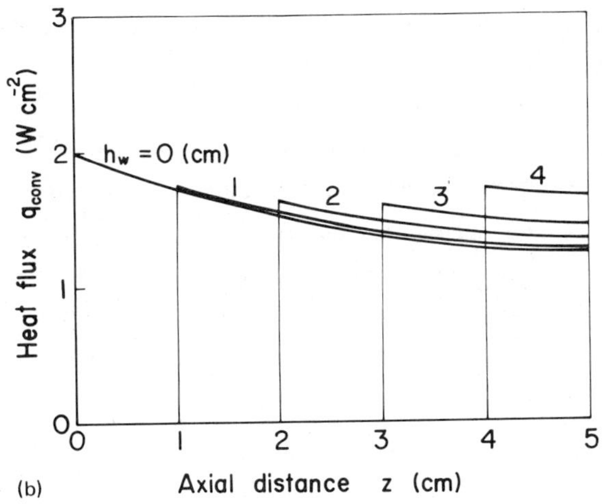

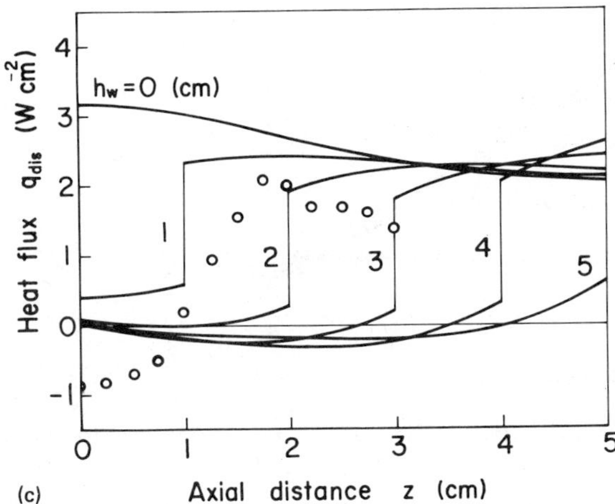

FIG. 44 (continued)

absorbed almost totally within the crucible for high wall height and
is dissipated outside the crucible. The heat dissipation from the
melt free surface is nearly the same as that in a vacuum.

   (i) Temperature distribution. The axial temperature distribu-
tion in the crystal for $\alpha_0 = 4 \times 10^{-4}$ W cm^{-2} K$^{-5/4}$ is also intermediate
between that for a vacuum and that for an ambient gas with constant
heat transfer coefficient. However, the axial temperature gradient
has features similar to those in a vacuum, as shown in Fig. 45. There
is a region where axial temperature gradient is nearly constant within
the crucible. This corresponds to the point at which heat is not
dissipated but absorbed. The axial temperature gradient at the interface is steep because the heat is dissipated by combined radiation and
gaseous convection.

   (ii) Interface shape. The interface shape for $\alpha_0 = 4 \times 10^{-4}$
W cm^{-2} K$^{-5/4}$ is shown in Fig. 46. As the wall height increases, the
interface shape suddenly changes from concave to convex. The sudden
change of the interface shape is due to sudden disappearance of the
heat dissipation by gaseous convection within the crucible, but is

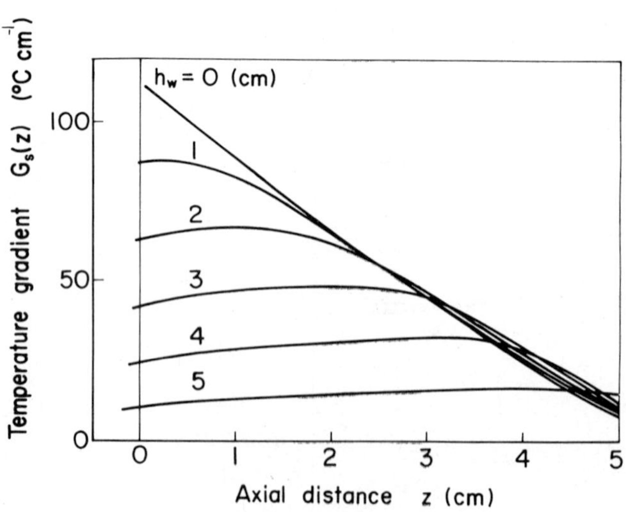

FIG. 45. Axial temperature gradients in the crystal for various values
of exposed crucible wall height. ($\alpha_0 = 4 \times 10^{-4}$ W cm^{-2} K$^{-5/4}$ and
$T_a = 300$ K.)

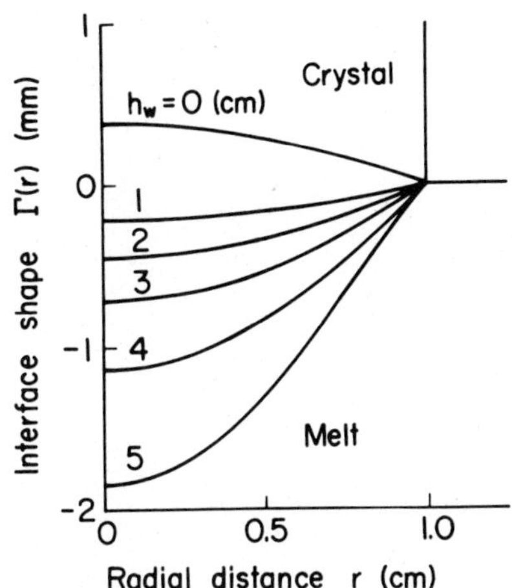

FIG. 46. Interface shapes for various values of exposed crucible wall height. ($\alpha_0 = 4 \times 10^{-4}$ W cm^{-2} K$^{-5/4}$ and $T_a = 300$ K.)

not observed in practice because the crucible is partly filled with melt. The deviations of the interface shape from planarity for various heat transfer coefficients are shown in Fig. 47. The interface shape reflects the nonuniformity of the heat flow from the melt

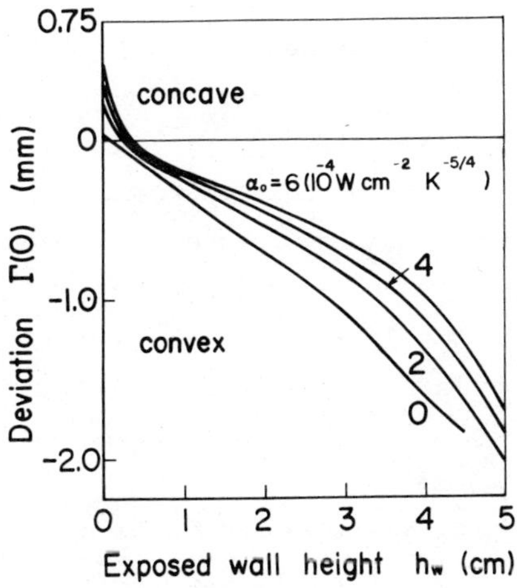

FIG. 47. Deviations of the interface shape from planarity as functions of exposed crucible wall height for various coefficients of heat transfer. ($T_a = 300$ K.)

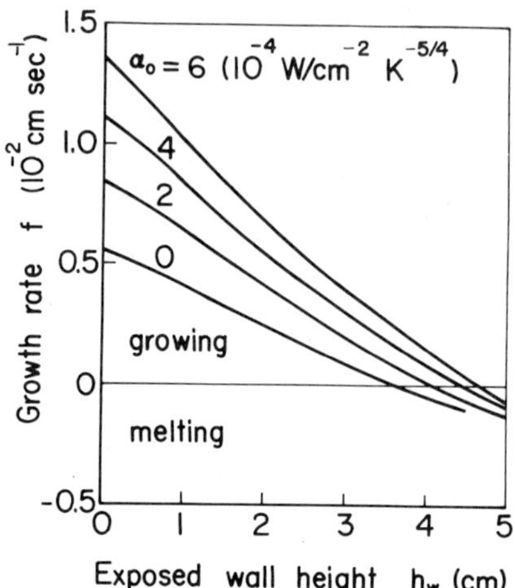

FIG. 48.   Growth rates as functions of exposed crucible wall height
for various coefficients of heat transfer.   (T_a = 300 K.)

to the interface because the heat flow in the crystal is axially uniform.   The growth rate for fixed crucible temperature is shown in
Fig. 48.   For large wall height, the crystal does not grow but rather
melts.   For continuous crystal growth, either the pulling rate or
the crucible temperature must be decreased with time.

C.   EXPERIMENTAL EVIDENCES

The temperature distribution in Ge was measured by Brice [17,72].
The temperature near the interface depended on the crystal radius at
$a^{1/2}(\partial T/\partial z) \simeq 105.5$ °C cm$^{-1/2}$ with $a(\partial^2 T/\partial z^2) \simeq 5.3$ °C cm^{-1}, and was
almost independent of the growth rate [17].   The heat flow was mainly
axial, suggesting that the radiative heat exchange between the crystal
and the crucible was effective and that the heat dissipation from the
crystal was relatively small within the crucible.   A detailed analysis
of the temperature distribution showed that the heat was absorbed
near the interface, as shown in Fig. 44.   The heat flux was calculated
from the temperature distribution by Eq. (37).

The temperature distribution in ZnWO$_4$ was measured by Scott [19].
The temperature gradient was constant within the crucible and decreased
exponentially outside the crucible.   The temperature distribution was
well expressed by

$$T = \begin{cases} \dfrac{T_m - (T_m - T_0)z}{z_0} & \text{for } z < z_0 \\[2em] (T_0 - T_a)\exp\left[\dfrac{-(2H)^{1/2}(z - z_0)}{a}\right] + T_a & \text{for } z > z_0 \end{cases} \tag{155}$$

where $T_m = 1190°C$, $T_0 = 1010°C$, $T_a = 425°C$, $z_0 = 0.75$ cm, $a = 0.7$ cm, and $H = 0.042$. This shows that the radiative heat dissipation was negligible within the crucible and that there was radial heat flow outside the crucible.

In growth of metals, radial heat flow within the crucible is usually neglected. Then the temperature gradient at the interface is controlled by the cross section of the crystal neck [73]:

$$\left(\frac{dT}{dz}\right)_{\text{interface}} = \left(\frac{a_{\text{neck}}}{a_{\text{interface}}}\right)^2 \left(\frac{dT}{dz}\right)_{\text{neck}} \tag{156}$$

These results show that the melt height markedly influences heat transfer through the exposed crucible wall and that the latter model is relevant. The exposed crucible wall then plays the role of an afterheater. Radiative heat exchange between the crystal and exposed crucible wall becomes more effective if the radius of the crystal is near that of the crucible.

The axial temperature gradient is decreased by an afterheater, by a heat reflector, and by a heat shield. On the other hand, the heat dissipation from the crystal surface may be controlled by blowing a gas jet across the crystal or the afterheater [74]. The axial temperature gradient then becomes steep. However, the effect of the gas jet on the temperature distribution is not effective unless the jet is directed to the region near the interface, because the crystal is cooled only the region where the jet is blown. Sometimes the gas-blowing equipment around the crystal actually prevents heat from being dissipated from the interface if the temperature of the gas is not sufficiently low [67].

The interface shape changes from concave to convex with increasing exposed crucible wall [18,68,70]. Strictly speaking, some facts

which are often thought as being effects of crystal length and melt
height are actually due to the effect of exposed crucible wall.

C. Interface Inversion

For high-Prandtl-number melts, the heat flow from the melt to the
interface sometimes changes during growth when the crystal diameter
exceeds a critical value and sudden remelt of the crystal occurs
[75-80]. At the same time, the fluid flow in the melt usually varies
from free convection to forced convection due to crystal rotation.
In diameter-controlled growth, abrupt inversion of the interface shape
from the extremely convex to planar occurs [75-80]. These facts are
explained by the transition of the fluid flow which causes the heat
flow from the melt to the interface to increase. Consequently, the
bulk melt temperature increases and the crystal begins to remelt.
Sometimes, the interface changes from concave to deep concave for
radiation-absorbing crystals [79].

For a large crucible, the transition of the fluid flow was not
observed with the interface inversion [78]. The flow before the inversion persists in the meantime, though it becomes weak. The flow
in the cylindrical region under the crystal mainly changes and it is
easily influenced by the input power. Then remelt and regrowth are
repeated with the flow change due to the input power variation. The
rapid regrowth induces the interface instability because of the constitutional supercooling if the melt height decreases.

The relations between the critical crystal diameter and the
crystal rotation are experimentally [76-80] and theoretically obtained
[56]. The critical crystal diameter is nearly linear for some cases
[76,78-80] and nearly proportional to the inverse square root of
the crystal rotation rate for others [77]. Strictly, the condition
of the interface shape inversion depends on the other factors such as
the crucible size, the insulation condition around the crystal, the
constitution of the melt, etc. Detailed observations are necessary.

If crystals are growing under a steep temperature gradient, the
interface shape is convex during the growth. If crystals are growing

under a shallow temperature gradient, the interface shape is concave during growth. The interface shape inversion is only observed under an intermediate temperature gradient [75].

For very-high-Prandtl-number melts, a fluid flow instability occurs when the crystal rotation rate exceeds a critical value [81-83]. At the critical rotation rate, the fluid flow changes from axially symmetric to asymmetric. Above the critical rotation rate, any stable crystal growth is impossible. The crystal diameter abruptly decreases and vanishes. On the melt free surface, some eddies are observed around the crystal. This type of hydrodynamic instability is quite similar to the baroclinic instability in the earth's troposphere [84].

D. Summary

To enable the experimentalist to obtain a planar interface, the various factors which influence interface shape are summarized here:

To make the interface shape less convex (more concave):

1. A large crystal radius is used.
2. The crystal rotation rate is increased.
3. The crucible rotation rate is reduced.
4. The growth rate is increased.
5. The crucible bottom is thermally insulated.
6. A flat-bottom crucible is used.

To make the interface shape less concave (more convex):

1. The crucible base is cooled.
2. A large crucible is used.
3. An afterheater is used.
4. A shallow melt with a high crucible wall is used.

VIII. CONTROL OF CRYSTAL GROWTH

A desired combination of crystal diameter and growth rate is usually controlled by the crucible temperature. For a given crucible

temperature, the crystal diameter may be determined by the growth
rate, which is controlled by the pulling speed. Control of the crystal diameter may easily be achieved if the time evolution of the
crystal diameter is predicted for a fixed growth rate. However, it
is a formidable task to predict the time evolution of the crystal
diameter during growth. Then the actual crystal growth is examined.
examined.

A.   Growth under Fixed Growth Rate and Crucible Temperature

The crystal diameter variations which are experimentally encountered
during growth are briefly described. The type of variation may
roughly be divided into two groups by the Biot number.

1.   *Low Biot Number*

Crystals of metals and semiconductors are included in this group.
Crystals with relatively constant diameter were easily grown [85,86].
The diameter variation is mainly due to the effect of the exposed
crucible wall [86]. At the initial stage of the growth, the crystal
diameter decreases with increasing crystal length because the radiative heat exchange between the crystal and the exposed crucible wall
increases. If the crystal top appears outside the crucible, the crystal diameter tends to increase because the heat dissipation from the
crystal top increases. Finally, the crystal diameter again decreases
with increasing crystal length because of the elevated melt temperature due to increasing exposed wall height. The degree of the diameter variation is larger for a larger crystal diameter. The final
wider section is very sensitive to the position of the heat source.
If the heat source is lifted, the wider section is increased. For
very low Biot number, the crystal diameter is independent of the
growth rate and of the crucible temperature if the bottleneck portion
is completely outside the crucible [85].

2.   *High Biot Number*

Crystals of oxides are included in this group. The crystal diameter
varies in an uncontrolled manner [71,74,87,88]. At the initial

stage of the growth, the diameter of those crystals rapidly increases with time. When the input power is rapidly decreased in order to decrease the crystal diameter, the crystal becomes disklike and sometimes separates from the melt. If the input power is decreased in order to rapidly increase the crystal diameter again, rapid radial growth occurs and the rotating crystal is sometimes broken by touching the crucible wall. This phenomenon is due to the crystal length dependence on the heat transfer from the crystal. If the Biot number is large, the heat is more easily dissipated from the disklike crystal than from the cylindrical crystal. Then the crystal prefers the disklike shape and grows radially outward.

B. Growth with Constant Crystal Diameter

Normally it is desired that a crystal grow with a constant diameter, except for the neck region. To obtain this, the growth rate and/or crucible temperature are usually varied.

The growth rate is generally determined by the heat balance at the interface, Eq. (134b):

$$f = \frac{1}{\rho_s \Delta H} (q_s - q_\ell) \tag{157}$$

where $q_s = -k_s(\partial T_s/\partial n)$ is the heat flux from the interface to the crystal and $q_\ell = -k_\ell(\partial T_\ell/\partial n)$ is the heat flux from the melt to the interface. For growth of a crystal with constant diameter and constant growth rate, as the crystal grows one has to steadily increase the heat transferred from the melt to the interface for a low Biot number and steadily decrease it for a high Biot number.

For example, if the one-dimensional model (Sec. V.A.1) with $H_e = H$ is effective, Eq. (157) becomes

$$f = \frac{1}{\rho_s \Delta H} \frac{k_s G_s(0) - k_\ell(T_b - T_m)}{\delta_f} \tag{158}$$

where $G_s(0) = (T_m - T_a)\beta(0)/a$, and $\beta(0)$ and δ_f are given in Eq. (43).

and Eq. (133) respectively. If it is assumed that the melt cannot be supercooled, then the maximum growth rate is attained when the melt is at the melting point or $q_\ell = 0$. The maximum growth rate is then given by

$$f_{max} = \frac{q_s}{\rho_s \, \Delta H} = \frac{1}{\rho_s \, \Delta H} \frac{k_s (T_m - T_a) \beta(0)}{a} \tag{159}$$

For a long crystal, Eq. (159) becomes

$$f_{max}^\infty = \frac{1}{\rho_s \, \Delta H} \frac{k_s (T_m - T_a)(2H)^{1/2}}{a} \tag{160}$$

For $H \gg 1$, Eq. (43) is approximated by

$$\beta(0) = (2H)^{1/2} \left\{ 1 + \left[\left(\frac{H}{2}\right)^{1/2} - 1 \right] \exp\left(\frac{-\ell}{\ell_0}\right) \right\} \tag{161}$$

where ℓ is the crystal length and $\ell_0 = (8H)^{-1/2} a$. In order to grow a crystal with a fixed growth rate, one has to decrease the bulk melt temperature during the growth. As the crystal grows, the bulk melt temperature is approximately given by

$$T_b = T_b^0 + C_1 \exp\left(\frac{-\ell}{\ell_0}\right) - C_2 \ell \tag{162}$$

where

$$T_b^0 = T_m + \frac{\delta_f^0}{k_\ell} (f_{max}^\infty - f) \rho_s \, \Delta H$$

$$C_1 = \frac{\delta_f^0}{k_\ell} \left[\left(\frac{H}{2}\right)^{1/2} - 1 \right] f_{max}^\infty \rho_s \, \Delta H$$

$$C_2 = \frac{(\rho_s/\rho_\ell)(a/b)^2 (T_b - T_m)}{d^0} \tag{163}$$

where f_{max}^∞ is the maximum growth rate for a long crystal, δ_f^0 is the initial thermal film thickness, and d^0 is the initial melt height.

The last term in Eq. (162) is due to the variation of the thermal film thickness, which is now assumed to be proportional to the melt height. An SBN crystal with a constant diameter is satisfactorily grown [71] by decreasing the bulk temperature similarly to Eq. (162). Practically, the constants T_b^0, C_1, and C_2 are semiempirically determined [71].

For $H \ll 1$, Eq. (43) is approximated by

$$\beta(0) = H\left[1 + \frac{2\ell}{a}\right] \tag{164}$$

To grow a crystal with a fixed growth rate, the bulk melt temperature has to be varied approximately as

$$T_b = T_b^0 + C_3 \ell \tag{165}$$

where

$$T_b^0 = T_m + \frac{\delta_f^0}{k_\ell}\left[\left(\frac{H}{2}\right)^{1/2} f_{max}^\infty - f\right] \rho_s \, \Delta H$$

$$C_3 = \frac{(\delta_f^0/k_\ell)\{(2H)^{1/2}(d^0/a) f_{max}^\infty - (\rho_s/\rho_\ell)(a/b)^2[(H/2)^{1/2} f_{max}^\infty - f]\}\rho_s \, \Delta H}{d^0} \tag{166}$$

The constant C_3 vanishes at a critical growth rate, where the crystal with a constant diameter is grown under a fixed crucible temperature. The critical growth rate is given by

$$f^c = \left[\frac{\rho_\ell}{\rho_s}\left(\frac{b}{a}\right)^2 \frac{d^0}{a} - \frac{1}{2}\right](2H)^{1/2} f_{max}^\infty \tag{167}$$

For a long crystal, the crystal radius change Δa due to the increase of the bulk melt temperature ΔT_b under a fixed growth rate is obtained from Eq. (158) by

$$\Delta a = -\frac{2k_\ell a^{3/2}}{k_s \delta_f (T_m - T_a)(2H/a)^{1/2}} \Delta T_b \tag{168}$$

This has been experimentally confirmed for low Biot number [17]. The

proportional constant is larger for a vacuum than for an ambient gas.
If the surrounding temperature is higher, the diameter change becomes
larger for a small decrease of the crucible temperature [89]. For a
fixed crucible temperature, the crystal diameter decreases with increasing growth rate as

$$\Delta a = - \frac{2\rho_s \, \Delta Ha^{3/2}}{k_s(T_m - T_a)(2h/a)^{1/2}} \, \Delta f \qquad (169)$$

C. Crystal Shape Instability

For a fixed crucible temperature, the growth rate may be varied with
exposed crucible wall to obtain a crystal with a constant diameter.
The growth rate often becomes zero for a certain wall height. To
avoid this situation, it is necessary to decrease the crucible temperature. Even if the crucible temperature is decreased in order to
keep a constant growth rate, the growth rate has sometimes to be decreased when a certain wall height is exceeded. This happens when
the heat from the melt to the interface becomes zero or when the temperature in the bulk melt approaches the melting point. Unless the
heat is forcibly removed from the crystal, the crystal growth stops.

In the crystal growth of oxides, the above situation sometimes
occurs and crystal shape instability accompanies it. Crystals with a
high Biot number have small thermal characteristic lengths. If the
crystal length within the crucible exceeds the thermal characteristic
length, it becomes difficult to dissipate heat through the crystal by
conduction. If the temperature in the melt is near the melting point,
the latent heat generated at the interface is mainly conducted through
the crystal and the growth rate becomes quite slow. Under this condition, a sudden deviation of the crystal shape from cylindrical is
sometimes observed [90]. This type of shape instability is due to
radiative heat exchange between the crystal and the other surfaces.
The radiative heat dissipation from the crystal surface is mainly
determined by the geometric factor f_{sa} between the crystal surface and
the surroundings, which is the lowest temperature among all surfaces.
The outward deviation of the crystal radius enhances the heat dissipation

through an increase of the factor f_{sa}. Contrarily, the inward deviation of the crystal radius suppresses the heat dissipation. Then the
deviation grows and stops when the radial heat balance is reattained
at the crystal-melt boundary by an increase or decrease in the radiative heat flow from the exposed crucible wall. To avoid this type of
instability, the crystal has to be grown under a condition such that
the factor f_{sa} is insensitive to a variation in the crystal shape during the growth. This can be achieved by using a large-diameter crucible and an afterheater which covers the crystal. With a large-
diameter crucible, the variation of the factor f_{sa} becomes relatively
small. With an afterheater or heat sheild, the heat is transferred
mainly by radiation from the crystal top surface and by gaseous convection if an ambient gas is used.

D. Temperature Fluctuation

Temperature and convective fluctuations in the melt induce solute inhomogeneities called striations in the crystal. Striations are un-
desirable for high-quality crystals. The most serious striations are
caused by crystal rotation in a nonuniform thermal environment. When
flow in the melt becomes strong, the convection becomes unstable and
then the fluid velocity and temperature fluctuate in time. Carruthers
[91,92] critically reviewed natural convection instabilities in melts.
A review on striations has been given by Carruthers and Witt [57].

The prime factor in distinguishing the stability of natural convection in the fluid is the Prandtl number. In high-Prandtl-number
melts, flow occurs primarily near the boundaries. In low-Prandtl-
number melts, the temperature is mainly governed by the boundary
conditions. The strength of natural convection in the melt is closely
related to the temperature difference between the crystal and the
crucible, as measured by the Grashof number $Gr_d = g\beta \, \Delta T d^3/\nu^2$. The
Rayleigh number $Ra_d = Gr_d Pr$ is usually used because there is a critical Rayleigh number below which the fluid flow is laminar and no temperature fluctuations occur in the fluid. The critical Rayleigh number Ra_d^c is estimates to be 2×10^4 in an ambient gas heated from below
and is smaller by an order of magnitude for typical melts [93].

1. *In the Ambient Gas*

In the gas around a growing crystal, the Rayleigh number normally is
larger than the critical one and turbulent flow is usually observed
[71,93,94]. The temperature fluctuations are very large, although
they decrease as one approaches the melt free surface. They are well
developed at positions several millimeters above the melt free surface [93]. Temperature fluctuations near the crystal are decreased
if the crystal is covered by a thermal shield. The temperature of
the melt near the melt free surface where a high destabilized temperature gradient is present is perturbed by temperature fluctuations
in the gas for oxide melts and fluctuates in phase because a convective
instability is induced [71,93,94].

2. *In the Melt*

Temperature fluctuations in the melt caused by natural convection may
be divided into two classes [94]. One is induced near the melt free
surface by a destabilizing temperature gradient. The other is induced
near the vertical melt-crucible boundary. Temperature fluctuations
which appear near the melt free surface are induced and enhanced by
those in the gas. A destabilizing temperature gradient is formed
by radiative heat dissipation from the melt free surface and becomes
large for high-temperature crystal growth. The temperature fluctuates
irregularly and transiently with a period of the order of a few seconds. For molten salts, the amplitude of the fluctuations is typically about 10°C. They increase with depth up to about a centimeter
below the melt free surface, and then decrease with further depths.
This type of temperature fluctuation may be suppressed by a baffle
in the melt [94].

Temperature fluctuations induced near the vertical melt-crucible
boundary are suppressed by decreasing the melt height [93]. The temperature fluctuates periodically with a period of several seconds.
For molten salts, the amplitude of these fluctuations is about 1°C.
This type of temperature fluctuation is suppressed by a horizontal
baffle because the instability arises from the vertical flow along
the crucible sides [94].

Crystal rotation induces growth rate fluctuations which cause severe rotational striations in the crystal [57]. The period of the fluctuation corresponds to the crystal rotation rate. For high crystal rotation rates at low growth rates, the fluctuations are so close that they may not be distinguishable and the magnitude of the fluctuations becomes small.

Crucible rotation also induces temperature fluctuations [94]. The period of the fluctuations corresponds to the crucible rotation rate. For ω_c = 16 rpm, the irregular temperature fluctuations override these and the amplitude of the fluctuations becomes large for molten oxides.

Combined crystal and crucible rotation induces complex temperature fluctuations in the melt [95].

E. Crystal Quality

The need for a high-quality crystal is sometimes inconsistent with other needs, such as obtaining a large crystal, growing with a high pulling rate, etc. Grown crystals are physically and chemically imperfect to some degree. The perfection of the crystal is essentially determined by heat and mass transfer during the crystal growth. However, their effects are not explicitly known. Detailed characterization of crystals is necessary for control and improvement of the crystal quality. Some of the physical imperfections are briefly examined here.

1. *Cracks*

Cracks are often introduced into oxide crystals [96], which are usually brittle during and after growth. To avoid cracks, the following methods are recommended:

1. The most anisotropic crystallographic axis is used for the growth direction.
2. The interface shape is forced to be planar by crystal rotation.
3. The thermal symmetry around the crystal is improved by stirring the melt.

4. Temperature gradients are decreased.
5. Crystals with circular cross sections are grown.

2. *Dislocations*

Dislocations may often be removed completely from a crystal if a long
thin neck is formed during the growth [97]. Once the dislocations no
longer intersect the growing interface, the crystal usually remains
dislocation-free in spite of changes in the thermal environment. Dislocations may be newly introduced when the crystal grows radially outward, with each dislocation being nearly perpendicular to the interface [78]. To suppress dislocation generation, the following methods
are recommended:

1. The thermal symmetry around the crystal is improved by stirring
    the melt or by use of heat pipes.
2. Thermal shock due to seeding or remelting is avoided.
3. Crystal necking is applied.
4. The temperature gradient is decreased by an afterheater or by
    increasing the ambient temperature.
5. Constitutional supercooling is avoided, by use of high purity and
    low growth rates.
6. The interface shape is forced to be planar, e.g., by crystal
    rotation.

3. *Thermal Stress*

Thermal stress in the crystal depends on the shapes of the isotherms
in the crystal, and is decreased if a planar interface is used. A
nearly strain-free crystal is obtained by making the interface planar
[60].

IX. DISCUSSION

There are many other factors to be considered in order to explain the
actual heat transfer. They are the temperature dependence of the
physical properties, the meniscus shape at the interface, the radiative heat transfer in the crystal and in the melt, the nonisothermal
crucible temperature, the Marangoni flow, the solutal flow, etc. Some
of these are now examined.

A. Temperature Dependence of Physical Properties

Many of the material properties are dependent on the temperature.
Presently, they are not known for almost all materials except a few
elemental ones over the wide range of temperature from room temperature to the melting point, and to still higher temperatures above the
melting point. Properties near the melting point are even lacking
in spite of their importance for crystal growth. Often these properties are simply estimated by crystal growers.

B. Meniscus Shape

Since the melt wets the crystal, a meniscus is pulled up above the
level of the bulk melt free surface. If a right cylinder is growing,
the meniscus height is approximately given by [98]

$$h_m = h_0 \left\{ \left[1 - \sin \theta_m + \left(\frac{h_0 \cos \theta_m}{4a} \right)^2 \right]^{1/2} - \frac{h_0 \cos \theta_m}{4a} \right\} \tag{170}$$

where $h_0 = (2\gamma/\rho_\ell g)^{1/2}$, γ is the surface tension, ρ_ℓ is the density, g
is the acceleration due to gravity, θ_m is the meniscus angle between
the vertical and the meniscus tangent at the solid-liquid interface,
and a is the crystal radius. As the meniscus height relative to the
crystal radius decreases with increasing crystal radius, the meniscus
height has less influence on the heat transfer. However, the thermal
effect of the meniscus is very important because the crystal diameter
is controlled through the meniscus shape [99]. Usually the meniscus
shape is varied by the heat flow through the position of the interface
relative to the bulk melt free surface. For diameter control, it is
necessary to investigate the meniscus in detail.

C. Effective Thermal Conductivity

For a nonopaque medium, radiative heat transfer occurs within it,
especially at elevated temperatures. In this case, Eqs. (6) and
(7) are no longer valid. The equation governing the temperature
becomes a tediously complex nonlinear integral-differential equation.

For the heat flow at steady state, some approximate methods for solution have been devised. Among them, the Rosseland approximation is very simple [100,101]. The thermal conductivity is replaced by an effective thermal conductivity:

$$k_{eff} = k + k_{rad} = k + \frac{16n^2\sigma T^3}{3\kappa} \tag{171}$$

where k_{rad} is the radiative thermal conductivity, n is the refractive index, and κ is the absorption coefficient. This expression is valid when the optical length $\tau_0 = \kappa L_c$ is larger than unity, where L_c is a characteristic length such as the crystal diameter, the crucible radius, or the melt depth. However, in the region within the thickness $1/\kappa$ of the boundary, this approximation is not valid.

The importance of the radiative heat transfer is effectively expressed by the conduction-radiation parameter (the Stark number) which represents the ratio of the heat transfer by conduction to that by radiation [101]:

$$N = \frac{k\kappa}{4\sigma T^3} \tag{172}$$

If the Stark number is large, heat transfer by conduction is dominant and Eqs. (6) and (7) are valid. If the Stark number is small, heat transfer by radiation is dominant and the effective thermal conductivity has to be used.

The Co-doped $ZnWO_4$ crystal has an absorption band between 1.0 and 2.2 µm and so radiative heat transfer occurs in the crystal. The effective thermal conductivity is proportional to the quantity of Co in the crystal and is given by

$$k_{eff} = 6.3C + 0.21 \tag{173}$$

where C is the concentration of Co and $k_s = 0.21$ W cm^{-1} K^{-1} is assumed [102].

The interface shape often becomes less convex if the purity of the crystal is decreased for oxide crystals [63]. In an impure crystal, radiation is more likely to be absorbed. Radiative heat transfer

in the crystal increases the effective thermal conductivity. Then
the Biot number becomes small for an impure crystal and the interface
shape becomes less convex. Similarly, a material which absorbs radiation has a less convex interface shape [61].

D. Nonisothermal Crucible Temperature

The temperature in the melt is usually controlled directly by varying
the input power to the crucible. The temperature of the crucible is
usually nonuniform and is the highest at the crucible side and lowest
at the crystal interface. The nonuniformity of the crucible temperature is closely related to the heating method employed. The temperature at the crucible bottom is generally relatively low because heat
is dissipated from the crucible bottom. The temperature at the melt
free surface is also low because of heat dissipation there. If the
temperature at the crucible bottom is relatively low, the interfacial
temperature gradient in the melt is decreased by crystal rotation
[60,61]. The temperature in the melt may be controlled by cooling
of the base of the crucible [88]. Such cooling decreases the interfacial temperature gradient in the melt. However, this control is
not effective for shallow melts.

In growth of oxide crystals, the crucible temperature varies
during growth by the change of the flow in the melt even if the input
power is kept constant [75-80]. In this case, it would be reasonable
to consider that the input power be specified.

E. Nonisothermal Interface Temperature

Actually, the interface temperature is not constant. It must be below
the melting point in order for the crystal to grow at a finite rate.
For many metals and semiconductors, the supercooling is often not
large unless there are facets on the interface. The facet is usually
observed at the central region on a convex interface, with the supercooling of the facet estimated to be a few degrees for semiconductors
[103,104]. If this supercooling is considered, the assumption that
the interface is an isotherm is not valid and the crystal growth
mechanism must be considered.

The facet often causes a core of slightly different properties in the crystal. In semiconductor crystals, the resistivity varies across such a cross section [58]. The resistivity on the facet is smaller than off facet. Generally, a core in a crystal is harmful to crystal quality and to potential application. A crystal without a core is usually desired and is usually obtained if the interface is planar or concave during growth. The faceted area of an originally dislocation-free crystal becomes smaller when dislocations are introduced into the crystal because of the decrease of the supercooling required for growth [104].

F. Growth Anisotropy

Crystals do not always grow isotropically. Crystals having high Jackson's α factors grow anisotropically [105]. The growth interface consists of two distinct types of the crystal surface: rough surface and smooth surface. The smooth surface is usually formed by a crystal plane with a low index. For example, a (111) facet is usually observed on the convex interface of Ge and Si, and (211) and (111) facets are observed on the convex interface of YAG. These facets may considerably influence the external shape of a crystal. The cross section is not always circular but may be formed by a roundly truncated polygon which has a crystal symmetry about the pulling axis [89, 106,107]. Such habits are markedly enhanced by growth in a small radial temperature gradient [106,107], at high rates, and with dislocation-free crystals.

Anisotropy of the thermal conductivity of the solid also produces a noncircular cross section. The cross section is an ellipse which is determined by the thermal conductivity tensor [4].

G. Encapsulated Liquid

For volatile melts, a liquid sheath of B_2O_3 and highly pressurized gas is used to prevent a component of the melt from evaporating into the atmosphere [108]. The liquid film of B_2O_3 thermally insulates

the melt from the atmosphere. The heat is radiated through the liquid
film. In the atmosphere, vigorous free convection occurs and the temperature fields in the melt and in the atmosphere may be highly homogenized [109].

If the crystal top is in the liquid film, the heat is transferred
mainly by radiation and the crystal diameter varies similarly to that
in a vacuum [110]. For a fixed growth condition, the crystal diameter
initially increases and then decreases by a self-limiting mechanism
due to the radiative heat exchange between the crystal and the exposed
crucible wall. If the crystal top appears outside the liquid film,
the heat is effectively transferred from the crystal top by vigorous
gaseous convection [110]. The crystal diameter varies according to
the diameter of the crystal just outside the liquid film. The angle
of a shoulder depends on the liquid film thickness. It is small for
a thick film because the heat conduction from the crystal seed becomes
ineffective and because the heat is transferred uniformly upward by
radiation.

X. CONCLUSIONS

Czochralski growth is performed with a lack of information on the
material properties and therefore important parts of the growth still
depend on the experience of the crystal growers. Presently, our understanding of heat transfer in crystal growth is qualitative rather
than quantitative because of this lack of information and because of
difficulties that arise when applying the fundamental equations of
heat transfer and the Navier-Stokes equation to a specific growth
process.

It is of great importance to measure the temperature fields in
the crystal, in the melt, and in the surroundings during crystal
growth in order to develop improved methods for control of growth
and for higher crystal quality. However, few attempts have been made
to obtain the temperature field because the conventional method using
thermocouples is harmful to the crystal being grown and often contami-
nates the furnace with the temperature-measuring materials. Thus it
is desired to measure the temperature nondestructively.

Appendix A.   THERMAL CONDUCTIVITY TENSOR

The thermal conductivity tensor for crystalline systems is as follows [4]:

1. Triclinic system:

$$\begin{pmatrix} k_{11} & k_{12} & k_{13} \\ k_{12} & k_{22} & k_{23} \\ k_{13} & k_{23} & k_{33} \end{pmatrix} \tag{173a}$$

2. Monoclinic system:

$$\begin{pmatrix} k_{11} & k_{12} & 0 \\ k_{12} & k_{22} & 0 \\ 0 & 0 & k_{33} \end{pmatrix} \tag{173b}$$

3. Orthorhombic system:

$$\begin{pmatrix} k_{11} & 0 & 0 \\ 0 & k_{22} & 0 \\ 0 & 0 & k_{33} \end{pmatrix} \tag{173c}$$

4. Tetragonal, trigonal, and hexagonal systems:

$$\begin{pmatrix} k_{11} & 0 & 0 \\ 0 & k_{11} & 0 \\ 0 & 0 & k_{33} \end{pmatrix} \tag{173d}$$

5. Cubic system:

$$\begin{pmatrix} k_{11} & 0 & 0 \\ 0 & k_{11} & 0 \\ 0 & 0 & k_{11} \end{pmatrix} \tag{173e}$$

Appendix B: RADIATIVE HEAT TRANSFER

A. Geometric View Factor

The geometric view factor from an infinitesimal surface area dA_i to the area dA_j is given by [8]

$$df_{dA_i - dA_j} = \cos \phi_i \, d\omega_{ij} \tag{174}$$

where ϕ_i is the angle between the normal direction to the surface dA_i and the direction from dA_i to dA_j and $d\omega_{ij}$ is the solid angle of the area dA_j viewed from the surface dA_i. Then the view factor $f_{dA_i - A_j}$ is given by

$$f_{dA_i - A_j} = \int_{A_j} \cos \phi_i \, d\omega_{ij} \tag{175}$$

Some of the view factors which appear in the Czochralski crystal growth heat transfer problem are summarized [111]. The symbols are $R = r/a$, $R'; = r'/a$, $Z = z/a$, $H_w = h_w/a$, and $L = \ell/a$.

1. *Factors Viewed from the Crystal Side Surface*

$a.$ *VIEW FACTOR TO THE MELT FREE SURFACE:*

$$df_{dA_s - dA\ell} = \frac{1}{\pi} \frac{\partial I(R, R_c, Z)}{\partial Z} RdR \tag{176}$$

$$f_{dA_s - A\ell} = \frac{R_c}{\pi(R_c^2 - 1)} J(R_c, Z) \tag{177}$$

where

$$I(R, R', Z) = \frac{\phi_m(R, R')}{2}$$
$$- \frac{(R^2 - R'^2 + Z^2)}{N_m(R, R', Z)} \tan^{-1} \left\{ \frac{N_m(R, R', Z) \tan[\phi_m(R,R')/2]}{(R - R')^2 + Z^2} \right\} \tag{178}$$

$$J(R, Z) = 2 \tan^{-1}\left[\frac{(R^2 - 1)^{1/2}}{Z}\right] - \frac{4Z(R^2 - 1)^{1/2}}{R^2 - 1 + Z^2} + Z\left[\frac{\pi}{2} - \sin^{-1}\left(\frac{1}{R}\right)\right]$$

$$+ \left[\frac{\Lambda(R,\ Z)Z}{\{\Lambda(R,\ Z)^2 - 1\}^{1/2}}\right]\left\{\frac{\pi}{2} + \sin^{-1}\left[\frac{R - \Lambda(R,\ Z)}{R\Lambda(R,\ Z) - 1}\right]\right\} \quad (179)$$

where

$$\phi_m(R,\ R') = \cos^{-1}\left(\frac{1}{R}\right) + \cos^{-1}\left(\frac{1}{R'}\right)$$

$$N_m(R,\ R',\ Z) = [(R + R')^2 + Z^2]^{1/2}\ [(R - R')^2 + Z^2]^{1/2}$$

$$\Lambda(R,\ Z) = \frac{R^2 + 1 + Z^2}{2R} \tag{180}$$

b. *VIEW FACTOR TO THE CRUCIBLE:*

$$f_{dA_s - A_c} = \begin{cases} \dfrac{1}{2} + \dfrac{R_c}{\pi(R_c^2 - 1)}\ J(R_c,\ H_w - L) & \text{for } Z < H_w \\[3ex] \dfrac{R_c}{\pi(R_c^2 - 1)}\ [J(R_c,\ Z - H_w) - J(R_c,\ Z)] & \text{for } Z > H_w \end{cases} \tag{181}$$

c. *VIEW FACTOR TO THE SURROUNDINGS:*

$$f_{dA_s - A_a} = 1 - (f_{dA_s - A_\ell} + f_{dA_s - A_c}) \tag{182}$$

2. *Factors Viewed from the Crystal Top Surface*

a. *VIEW FACTOR TO THE CRUCIBLE:*

$$f_{dA_e - A_c} = \begin{cases} \dfrac{1}{2}\left[1 + \dfrac{R^2 - R_c^2 + (H_w - L)^2}{N_m(R,\ R_c,\ H_w - Z)}\right] & \text{for } L < H_w \\[3ex] 0 & \text{for } L > H_w \end{cases} \tag{183}$$

b. *VIEW FACTOR TO THE SURROUNDINGS:*

$$f_{dA_e - A_a} = 1 - f_{dA_e - A_c} \tag{184}$$

3. Factors Viewed from the Melt Free Surface

a. VIEW FACTOR TO THE CRYSTAL SIDE SURFACE:

$$df_{A_\ell - dA_s} = \frac{1}{\pi} \frac{\partial I(R,\ R_c,\ Z)}{\partial Z}\ dZ \tag{185}$$

$$f_{dA_\ell - A_s} = \frac{1}{\pi}\left[\sin^{-1}\left(\frac{1}{R}\right) + I(R,\ 1,\ L)\right] \tag{186}$$

b. VIEW FACTOR TO THE SURROUNDINGS:

$$f_{dA_\ell - A_a} = \frac{1}{\pi}[I(R,\ R_c,\ H_w) - I(R,\ 1,\ L)] \quad \text{for } L > H_w \tag{187}$$

c. VIEW FACTOR TO THE CRUCIBLE:

$$f_{dA_\ell - A_c} = 1 - \frac{1}{\pi}\left[\sin^{-1}\left(\frac{1}{R}\right) + I(R,\ R_c,\ H_w)\right] \quad \text{for } L > H_w \tag{188}$$

4. Factor Viewed from the Crucible Side Surface

a. VIEW FACTOR TO THE CRUCIBLE ITSELF:

$$f_{A_c - A_c} = 1 - \frac{2}{\pi R_c}\left[\frac{\pi}{2} - \cos^{-1} N_n(R_c,\ H_w)\right] + \frac{2H_w}{\pi R_c}\left[\frac{\pi}{2} - \sin^{-1}\left(1 - \frac{1}{R_c^2}\right)\right]$$

$$+ \frac{2(4R_c^2 + H_w)^{1/2}}{\pi R_c}\left\{\frac{\pi}{2} + \sin^{-1}\left[1 - \frac{2}{R_c^2} N_n(R_c,\ H_w)^2\right]\right\} \tag{189}$$

where

$$N_n(R,\ Z) = \frac{Z}{[4(R^2 - 1) + Z^2]^{1/2}} \tag{190}$$

B. Radiative Heat Dissipation

The radiative heat dissipation from the surface dA_i is approximately given by

$$q_{rad} = \varepsilon_i \sigma \sum_{j=1}^{N} c_j f_{dA_i - A_j} (T_i^4 - T_j^4) \tag{191}$$

The heat dissipation from various surfaces are summarized.

a.   CRYSTAL SIDE SURFACE:

$$q_{rad,s} = \varepsilon_s \sigma [\varepsilon_\ell f_{s\ell} (T^4 - T_\ell^4) + \varepsilon_c f_{sc} (T^4 - T_c^4) + f_{sa} (T^4 - T_a^4)]$$

$$(192)$$

If the melt temperature is not homogeneous, $\bar{T}_\ell$ is used in place of T_ℓ:

$$\bar{T}_\ell^4 = \int_A T_\ell^4 \, df_{s1}$$

$$(193)$$

b.   CRYSTAL TOP SURFACE:

$$q_{rad,e} = \varepsilon_s \sigma [\varepsilon_c (1 - f_{ea}) (T^4 - T_c^4) + f_{ea} (T^4 - T_a^4)] \qquad (194)$$

c.   MELT FREE SURFACE:

$$q_{rad,\ell} = \varepsilon_\ell \sigma [\varepsilon_s f_{\ell s} (T^4 - T_s^4) + \varepsilon_c f_{\ell c} (T^4 - T_c^4) + f_{\ell a} (T^4 - T_a^4)]$$

$$(195)$$

If the crystal temperature is not homogeneous, $\bar{T}_s$ is used in place of T_s:

$$\bar{T}_s = \int_{A_s} T_s^4 \, df_{\ell s}$$

$$(196)$$

Appendix C.   GROWTH LAW

The generalized growth laws for the different known mechanisms may be
written by using the velocity of the growth normal to the surface f,
the interfacial supercooling required to drive the attachment process, ΔT, and kinetic constants μ_i. For a single component system [9]:

1.   Uniform interface attachment:

$$f = \mu_0 \, \Delta T \qquad (197a)$$

2.   Large-edge motion:

$$f = \mu_\infty \, \Delta T \qquad (197b)$$

3.   Screw-dislocation source:

$$f = \mu_1 \; \Delta T \left| \Delta T \right|$$ (197c)

4.   Two-dimensional nucleation (layer-passage limited):

$$f = \mu_4 \; \exp\left(\frac{\mu_3}{3 \; \Delta T}\right)$$ (197d)

5.   Two-dimensional nucleation (layer-source limited):

$$f = b_0 A \mu_2 \; \exp\left(\frac{\mu_3}{\Delta T}\right)$$ (197e)

where b_0 is the molecular step height normal to the interface and A is the area of the crystal face.

6.   Twin-plane reentrant corner nucleation:

$$f = \mu_5 \; \exp\left(\frac{\mu_6}{\Delta T}\right)$$ (197f)

Appendix D. NUMERICAL METHOD

The numerical method is now briefly described. The desired quantities are the temperature field, the solute concentration field, the velocity field in the melt, and the interface shape. The temperature field and the solute concentration field are readily obtained by solving the partial differential equations if the velocity field and the interface shape are specified. The velocity field is not easily obtained if the Reynolds number or the Grashof number is large and various numerical methods are devised to solve the Navier-Stokes equation. There is no general method for determining the interface shape, although it must satisfy the interfacial heat balance and be at the melting point. The relaxation method [112] is sometimes used to obtain the solutions of differential equations. Practically, all of these unknowns influence one another and the procedure for solving them is complicated and tedious. The essentials of the procedure are as follows.

A. Derivation of Difference Equation

The physical region in the (r, z) plane is divided into mesh with spacings of Δr and Δz, as shown in Fig. 49. At each cross-point of the mesh, next quantities are defined:

$$r_i = i \, \Delta r \tag{198}$$

$$z_j = j \, \Delta z \tag{199}$$

$$P(i,j) = P(r_i, z_j) \tag{200}$$

$$\nabla_r P(i, j) = \frac{P(i + 1, j) - P(i - 1, j)}{2 \, \Delta r} \tag{201}$$

$$\nabla_r^2 P(i, j) = \frac{P(i + 1, j) - 2P(i, j) + P(i - 1, j)}{\Delta r^2} \tag{202}$$

$$\nabla_z P(i, j) = \frac{P(i, j + 1) - P(i, j - 1)}{2 \, \Delta z} \tag{203}$$

$$\nabla_z^2 P(i, j) = \frac{P(i, j + 1) - 2P(i, j) + P(i, j - 1)}{\Delta z^2} \tag{204}$$

In the finite difference approximation, the partial differential operators, $\partial/\partial r$, $\partial^2/\partial r^2$, $\partial/\partial z$, and $\partial^2/\partial z^2$, in a differential equation

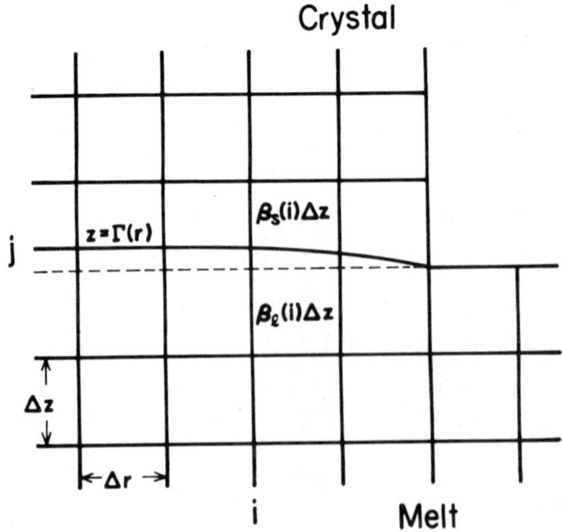

FIG. 49. Mesh near the interface. Irregular mesh spacings are used near the interface.

are replaced by the difference operators, ∇_r, ∇_r^2, ∇_z, and ∇_z^2, respectively. At the boundary, the definitions (201) and (203) are not always appropriate and the definition below is used for $\nabla_r P(i, j)$:

$$\nabla_r P(i, j) = \frac{\mp [3P(i, j) - 4P(i\pm1, j) + P(i \pm 2, j)]}{2 \, \nabla r} \tag{205}$$

where upper signs are used at the lower boundary and lower signs are at the upper boundary. Similarly, $\nabla_z P(i, j)$ may be defined.

At the points of the nearest neighbor of the interface, the definitions (203) and (204) are not always appropriate again because of the irregularity in mesh pattern due to nonplanarity of the interface shape. In this case, the operators ∇_z and ∇_z^2 are defined as follows:

$$\nabla_z P(i, j) = \left[\frac{\beta_i'}{\beta_i(\beta_i + \beta_i')} P(i, j + 1) + \frac{\beta_i - \beta_i'}{\beta_i \beta_i'} P(i, j) \right.$$
$$\left. - \frac{\beta_i}{\beta_i'(\beta_i + \beta_i')} P(i, j - 1) \right] \Delta z \tag{206}$$

$$\nabla_z^2 P(i, j) = 2\left[\frac{P(i, j + 1)}{\beta_i(\beta_i + \beta_i')} - \frac{P(i, j)}{\beta_i \beta_i'} + \frac{P(i, j - 1)}{\beta_i'(\beta_i + \beta_i')} \right] /\Delta z^2 \tag{207}$$

where $\beta_i \, \Delta z$ and $\beta_i' \, \Delta z$ are the distances, respectively between the points (i, j) and $(i, j + 1)$ and between (i, j) and $(i, j - 1)$. $\beta_i = 1$ and $\beta_i' = \beta_s(i)$ if the point is in the crystal and $\beta_i = \beta(i)$ and $\beta_i' = 1$ if it is in the melt.

At the points on the interface, the operators ∇_r and ∇_z are not appropriate again and are defined respectively by

$$\nabla_r P(i, j) = \frac{\mp [3P(i, j) - 4P*(i \pm 1, j) + P*(i \pm 2, j)]}{2 \, \Delta r} \tag{208}$$

$$\nabla_z P(i, j) = \frac{\pm[(1 + 2B_\pm)P(i, j) - (1 + B_\pm)^2 P(i, j \pm 1) + B_\pm^2 P(i, j \pm 2)]}{B_\pm (1 + B_\pm)\Delta z} \tag{209}$$

where

$$P^*(i + k, j) = P(i + k, j \pm 1) + \nabla_z P(i + k, j \pm 1)\beta_\pm$$

$$+ \frac{1}{2} \nabla_z^2 P(i + k, j \pm 1) \beta_\pm^2 \quad \text{for } k = \pm 1, \pm 2 \quad (210)$$

and where $\beta_+ = \beta_s(i + k)$, $\beta_- = \beta_\ell(i + k)$. At the next points of the interface edge, the definition (208) is not always defined and the following definition is used:

$$\nabla_r P(i, j) = \frac{\mp[P(i, j) - P^*(i \pm 1, j)]}{\Delta r} \quad (211)$$

B. Solution of Differential Equation

The differential equations to be integrated are classified into three groups:

$$\frac{1}{r} \frac{\partial}{\partial r} \left(r \frac{\partial S}{\partial r} \right) + \frac{\partial^2 S}{\partial z^2} = R \quad \text{for } T \quad (212)$$

$$\frac{\partial}{\partial r} \left[\frac{1}{r} \frac{\partial}{\partial r} (rS) \right] + \frac{\partial^2 S}{\partial z^2} = R \quad \text{for } v \text{ and } \chi \quad (213)$$

$$\frac{\partial^2 S}{\partial r^2} - \frac{1}{r} \frac{\partial S}{\partial r} + \frac{\partial^2 S}{\partial z^2} = R \quad \text{for } \psi \quad (214)$$

where R means the remaining terms. A method to integrate these equations numerically is now described. Equation (212) may be taken as an example of the three because the other equations are similarly integrated with slight modifications.

The difference equation corresponding to Eq. (212) is expressed by

$$S(i, j) = \frac{1}{4} \left[\left(1 + \frac{1}{2i}\right) S(i + 1, j) + \left(1 - \frac{1}{2i}\right) S(i - 1, j) \right.$$

$$\left. + S(i, j + 1) + S(i, j - 1) - R(i, j) \right] \quad (215)$$

Equation (215) may be solved by successive iteration, i.e., the νth approximate value of $S(i, j)$ is obtained by

$$S^{(\nu)}(i, j) = \frac{1}{4}\left[\left(1 + \frac{1}{2i}\right) S^{(\nu')}(i + 1, j) + \left(1 - \frac{1}{2i}\right) S^{(\nu')}(i - 1, j)\right.$$

$$\left. + S^{(\nu')}(i, j + 1) + S^{(\nu')}(i, j - 1) - R^{(\nu')}(i, j)\right] \tag{216}$$

where ν' is ν or $(\nu - 1)$ and $R^{(\nu')}(i, j)$ is the newest obtained value. Similarly, the νth approximate value of $S(i, j)$ at the points of the boundaries and at the points near the interface are obtained. Then these iterations are repeated until the following convergent condition is satisfied:

$$\frac{\left|S^{(\nu)}(i, j) - S^{(\nu - 1)}(i, j)\right|}{\max_{i,j}\left|S^{(\nu)}(i, j)\right|} < \varepsilon_r \tag{217}$$

where ε_r is the cutoff value.

C. Determination of Interface Shape

To obtain the interface shape $z = \Gamma(r)$, the next quantities are defined:

$$Q_s = -k_s\left(\frac{\partial T_s}{\partial n}\right) = -k_s\left(\frac{\partial T_s}{\partial z}\right)(1 + m_i^2)^{1/2} \tag{218}$$

$$Q_\ell = -k_\ell\left(\frac{\partial T_\ell}{\partial n}\right) = -k_\ell\left(\frac{\partial T_\ell}{\partial z}\right)(1 + m_i^2)^{1/2} \tag{219}$$

$$Q_f = \frac{\rho_\ell\,\Delta H f}{(1 + m_i^2)^{1/2}} \tag{220}$$

$$Q_d = Q_s - Q_\ell \tag{221}$$

where m_i is the slope of the interface defined by

$$m_i = \frac{d\Gamma}{dr} \tag{222}$$

The condition of heat balance at the interface is now given by

$$Q_d = Q_f \tag{223}$$

In the finite difference approximation, Eq. (222) is expressed by

$$m_i(i) = m_i(r_i) = \frac{[\beta_\ell(i + 1) - \beta_\ell(i - 1)]\,\Delta z}{2\,\Delta r} \tag{224}$$

At the edge of the interface, the next conditions hold:

$$m_i(0) = 0 \tag{225a}$$

$$m_i(i_p) = -\frac{\partial T_s/\partial r}{\partial T_s/\partial z} \tag{225b}$$

For a given interface shape $z = \Gamma_0(r)$, the quantities (218) – (222) may be calculated by using the solutions of the differential equations. With these quantities, an approximate interface shape may be obtained as follows. First, an approximate growth rate is calculated from Eq. (223) at the edge of the interface, where the position of the interface is assumed to be zero, and $Q_f(i)$ is calculated by using Eq. (220). Usually, Eq. (223) is not satisfied over the interface, i.e., $Q_d(i)$ is not always equal to $Q_f(i)$. Second, the interface shape is deformed in order to make $Q_d(i)$ equal to $Q_f(i)$. Because the heat flow at the interface is mainly axial, the amount of the deformation of $\beta_\ell(i)$ is given by

$$\Delta\beta_\ell(i) = \frac{Q_f(i) - Q_d(i)}{Q_s(i) + Q_\ell(i)} \tag{226}$$

Third, this iteration is repeated until the νth approximate value converges in the sense of Eq. (217). Finally, the interface is given by

$$z = \Gamma(r) = [\beta_\ell(r) - 1]\,\Delta z \tag{227}$$

D. Solutions of Simultaneous Differential Equations

The solution of each differential equation and the interface shape are obtainable if the other unknowns are specified. However, these solutions are not self-consistent because they do not always satisfy the differential equations simultaneously. To obtain the self-consistent solutions, the following outer-iteration is performed. First, for seven unknowns T, u, v, w, χ, ψ, and Γ, we set initial values T_0, u_0, v_0, w_0, χ_0, ψ_0, and Γ_0, respectively. Usually, as the initial values the solutions of the previous problem with slightly different parameters and the planar interface shape are used. Second, we obtain a new approximate solution T_0' of T with the method described in Sec. B, using these values. Then, the mean value $T_1 = (T_0 + T_0')/2$ is used in place of T_0. Similarly, new approximate values v_1, χ_1, and ψ_1 are obtained, step by step. u_1 and w_1 are calculated from Eq. (97). A new approximate interface shape Γ_1 is obtained with the method described in Sec. C. Now, all of the first approximate values T_1, u_1, v_1, w_1, χ_1, ψ_1, and Γ_1 are obtained. Finally, the above outer-iteration is repeated until all of the solutions satisfy the condition (217) simultaneously.

In this procedure, the growth rate is determined from the heat balance at the interface. If the growth rate is specified initially, more computational time is required because the crucible temperature is varied to adjust the growth rate. This step disturbs the temperature and velocity fields in the melt considerably.

The numerical results shown in Sec. VII.B. are obtained with the method described above for $\Delta r = \Delta z = 0.25$ (cm). If half-mesh spacings are used, the interface shape is slightly changed while the velocity field is considerably altered. This method is applicable under the limited condition that the interface shape is nearly planar and the flow in the melt is weak. For other conditions, it is necessary to improve this method.

ACKNOWLEDGMENTS

The author is grateful to several anonymous persons for reviews of the manuscripts. He is especially indebted to Professor W. R. Wilcox of Clarkson College, who makes this text possible, for much valuable criticism and improvement of English.

NOMENCLATURE

a Crystal radius, cm

b Crucible radius, cm

C Solute concentration, mol cm^{-3}

C_b Solute concentration of bulk fluid

C_i Constant, i = 0, 1, 2, 3

c_i Constant, i = 0, 1, 2

c_p Specific heat at constant pressure, J g^{-1} K^{-1}

D Diffusion coefficient, cm^2 sec^{-1}

D Dimensionless melt height, d/a

d Melt height, cm

F Function defined by Eq. (101)

F_p Dimensionless pulling rate, $f_p/(a\omega_s)$

f Growth rate, cm sec^{-1}

f_{ij} Geometric factor viewed from ith surface to jth surface

f_p Pulling rate, cm sec^{-1}

f_0 Growth rate for planar interface, cm sec^{-1}

G Temperature gradient, °C cm^{-1}

G Function defined by Eq. (101)

Gr Grashof number, $g\beta \, \Delta T L_c^3 \, \nu^2$

Gr_h Grashof number, $g\beta \, \Delta T b^3/\nu^2$

g Acceleration due to gravity, cm sec^{-2}

$\vec{g}$ Acceleration vector due to gravity

H Biot number, $h_a a/k_s$ for convective heat transfer, $\varepsilon_s \sigma T_m^3 a/k_s$ for radiative heat transfer, and $(\varepsilon_s \sigma T_m^3 a + \alpha_h T_m^{1/4})/k_s$ for combined convective and radiative heat transfer

H Function defined by Eq. (101)

H_f Biot number for finite crystal length, $H(T_m/\Delta T_e)$

H_m Biot number of melt, $(\varepsilon_\ell \sigma T_m^3 a/k_\ell)(T_m/\Delta T_c)$

H_w Dimensionless exposed crucible wall height, h_w/a

H Dimensionless interfacial velocity, $-(\rho_\ell/\rho_s)f/(\nu\omega_s)^{1/2}$

h Heat transfer coefficient, W cm^{-2} K^{-1}

h_m Meniscus height, cm

h_w Exposed crucible wall height, cm

h_0 Capillary constant, $(2\gamma/\rho_\ell g)^{1/2}$, cm

K Curvature, cm

k Thermal conductivity, $W\ cm^{-1}\ K^{-1}$

k_{ij} Element of Thermal conductivity tensor

k_m Thermal conductivity of solid at melting point

k_m Mass transfer coefficient, $cm\ sec^{-1}$

k Equilibrium segregation coefficient

L Dimensionless crystal length, ℓ/a

L_c Characteristic length, cm

L_h Ratio of latent heat to dissipated heat, $f\rho_s\ \Delta H/[h_a(T_m - T_a)]$

L_p Periphery of crystal cross section, cm

ℓ Crystal length, cm

Ma Marangoni number, $(-\partial\gamma/\partial T)\ \Delta T L_c/(\alpha_t\mu)$

m Slope of liquidus line, $K\ cm^{+3}\ mol^{-1}$

N Stark number, $k\kappa/(\varepsilon\sigma T^3)$

N_m Mass flux, $mol\ sec^{-1}\ cm^{-2}$

N Diffusivity coefficient, $\nu(U_c/L_c)/(c_p\ \Delta T)$

n Refractive index

n Constant, 4 for laminar flow and 3 for turbulent flow

$\vec{n}$ Unit vector normal to surface (n_1, n_2, n_3)

P Dimensionless pressure, $(p/\rho_\ell)/(\Omega_c U_c L_c)$

Pr Prandtl number, ν/α_t'

p Pressure, $N\ cm^{-2}$

Q Total heat flux, W

Q_0 Dimensionless heat flux, $q_0 a/(k_s T_m)$

q Heat flux, $W\ cm^{-2}$

q_0 Specified heat flux

$\vec{q}$ Heat flux vector (q_x, q_y, q_z)

R Dimensionless radial position, r/a

Ra Rayleigh number, $g\beta\ \Delta T L_c^3/(\alpha_t \nu)$

Re Reynolds number, $L_c U_c/\nu$

Re_m Mean Reynolds number, $(1 + s)Re_r$

Ro Rossby number, $L_c \Omega_c/U_c$

r Radial position, cm

S Cross-sectional area of crystal, cm

S_h Ratio of transferred heat to dissipated heat, $h_\ell (T_b - T_m)$ $/[h_a (T_m - T_a)]$

s Rotation ratio

T Temperature, $°C$ or K

T_b Bulk Melt Temperature

T_m Melting point of pure material

T_0 Temperature at axial point, $z = z_0$

t Time, sec

U Dimensionless radial valocity, u/U_c

U_c Characteristic velocity, cm sec^{-1}

u Radial velocity, cm sec^{-1}

V Dimensionless azimuthal velocity, v/U_c

V Volume, cm^3

v Velocity, cm sec^{-1}

v Aximuthal velocity

$\vec{V}$ Velocity vector (v_x, v_y, v_z)

W Dimensionless axial velocity, w/U_c

w Axial velocity, cm sec^{-1}

$\vec{X}$ Body force, N cm^{-3}

x Rectangular coordinate, cm

y Rectangular coordinate, cm

Z Dimensionless axial position, z/a

z Rectangular coordinate, cm

z Axial position, cm

z Distance from interface, cm

α Péclet number, $f\rho_s c_{ps} a/k_s$

α_h Heat transfer coefficient defined by Eq. (152), W cm^{-2} $\text{K}^{-5/4}$

α_n nth Root of $\alpha J_1(\alpha) - HJ_0(\alpha) = 0$

α_t Thermal diffusivity, $k/(\rho c_p)$, cm sec^{-1}

α_0 Specified heat transfer coefficient, W cm^{-2} $\text{K}^{-5/4}$

β Thermal expansion coefficient, $-(\partial \ln V/\partial T)_C$, K^{-1}

β Dimensionless temperature gradient in solid, $aG_s/\Delta T$, where $\Delta T = T_m - T_a$ for convective heat dissipation and $\Delta T = T_m$ for radiative heat dissipation

β_m Concentration coefficient of volumetric expansion, $-(\partial \ln V/\partial C)_T$, $(\text{mol cm}^{-3})^{-1}$

Γ Interface shape, $z = \Gamma(r)$, cm

Γ_m Mean deviation from planar interface, V/S, cm

γ Surface tension, N cm^{-1}, or interfacial energy, J cm^{-2}

Δ Function defined by Eq. (128)

ΔH Latent heat, J g^{-1}

ΔT Characteristic temperature, $^\circ$C or K

ΔT Interfacial supercooling, $^\circ$C

ΔT_b Increment of bulk melt temperature, $^\circ$C

ΔS Entropy of fusion, $\text{J g}^{-1}\text{ K}^{-1}$

Δa Increment of crystal diameter, cm

Δf Increment of growth rate, cm sec^{-1}

Δ_t Dimensionless thermal layer thickness, $\delta_t(\omega_s/\nu)^{1/2}$

$\Delta\omega$ Rotation difference, $\omega_s - \omega_c$, rad sec^{-1}

δ Boundary layer thickness, $(\nu/\Omega_c)^{1/2}$, cm

δ_f Thermal film thickness, $\Delta T/G_\ell(0)$, cm

δ_{ij} Cronecker's delta

δ_t Thermal layer thickness, cm

ε Emissivity

ζ Dimensionless axial distance, $z(\omega_s/\nu)^{1/2}$

η Radiation efficiency defined by Eq. (153)

θ Dimensionless temperature, $T/\Delta T$ or $(T - T_a)/\Delta T$

θ_b Angle between interface and vertical crystal boundary, rad

θ_m Meniscus angle between vertical and meniscus at solid-liquid interface, rad

κ Absorption coefficient, cm^{-1}

μ Viscosity, N sec cm^{-2}

ν Kinematic viscosity, μ/ρ_ℓ, $\text{cm}^2\text{ sec}^{-1}$

ξ Dimensionless axial distance, z/d

ξ_0 Dimensionless distance from interface to dividing surface

ρ Density

μ Dimensionless radial distance, $r(\omega_s/\nu)^{1/2}$

σ Stefan-Boltzmann constant, $5.67 \times 10^{-12}\text{ W cm}^{-2}\text{ K}^{-4}$

σ_{ij} Element of stress tensor, N cm^{-2}

τ Dimensionless time, $\Omega_c t$

τ_0 Optical length, κL_c

Φ Dimensionless dissipation function, $(L_c/U_c)^2 \phi$

ϕ Dissipation function, sec^{-2}

χ Vorticity function defined by Eq. (96)

ψ Stream function defined by Eq. (97)

Ω_c Characteristic frequency, sec^{-1}

ω Rotation rate, rad sec^{-1}

Subscripts

a   Atmosphere, gas or surroundings

c   Crucible

cond Conduction

conv Convection

d   Based on $L_c = d$

dis   Dissipation

e   Cold end

eff   Effective

eq   Equilibrium

i   ith Surface

in   Input

ℓ Liquid

max   Maximum

n   Normal to surface

r   Based on $L_c = a$

s   Solid or crystal

x   x-Component

y   y-Component

z   z-Component

1,2, Component of rectangular coordinate, 1 for x, 2 for y, and 3 for
  3   z

Superscripts

c   Critical

0   Initial

∞ For long crystal

'   Based on $\Omega_c = \omega_c$

Overline

—   Averaged

REFERENCES

1. T. B. Reed, in *Crystal Growth* (Peiser, ed.), Pergamon, New York, 1967, p. 39.

2. B. Cockayne, M. Chesswas, D. J. Born, and J. D. Filby, *J. Mater. Sci.*, *4*, 236 (1969).

3. T. B. Reed and E. R. Pollard, *J. Crystal Growth*, *2*, 243 (1968).

4. H. S. Carslaw and J. C. Jeager, *Conduction of Heat in Solids*, Clarendon, Oxford, 1959, chap. 1.

5. R. B. Bird, W. E. Stewart, and E. N. Lightfood, *Transport Phenomena*, Wiley, New York, 1960, chap. 19.

6. W. H. McAdams, *Heat Transmission*, McGraw-Hill, New York, 1954, chap. 7.

7. W. R. Wilcox, in *Preparation and Properties of Solid State Materials*, Vol. 1 (R. A. Lefever, ed.), Marcel Dekker, New York, 1971, p. 37.

8. L. O. Landau and E. M. Lifshitz, *Fluid Mechanics*, Pergamon, London, 1959, chap. 5.

9. W. A. Tiller, in *Solidification*, American Society for Metals, Metals Park, 1971, chap. 3.

10. J. C. Baker and J. W. Cahn, in *Solidification*, American Society for Metals, Metals Park, 1971, chap. 2.

11. V. H. S. Kuo and W. R. Wilcox, *J. Crystal Growth*, *12*, 191 (1972).

12. W. R. Wilcox and R. L. Duty, *J. Heat Transfer, Trans. ASME*, *88*, 45 (1966).

13. V. H. Gol'dfarb, B. M. Tol'tsman, An. V. Donski, and A. V. Stepanov, *Sov. Phys.-Crystallog.*, *10*, 450 (1966).

14. N. Kobayashi, abstracts of the 35th Autumn Meeting of the Japanese Society of Applied Physics, Vol. 2, p. 150.

15. K. V. Ravi, *J. Crystal Growth*, *39*, 1 (1977).

16. D. T. J. Hurle, *Prog. Mater. Sci.*, *10*, 79 (1962).

17. J. C. Brice, *J. Crystal Growth*, *2*, 395 (1968).

18. E. Billing, *Proc. Roy. Soc. A235*, 37 (1955).

19. R. A. M. Scott, *J. Crystal Growth*, *10*, 39 (1971).

20. J. A. Burton, R. C. Prim, and W. P. Slichter, *J. Chem. Phys.*, *21*, 1987 (1953).

21. W. G. Cochran, *Proc. Camb. Phil. Soc.*, *30*, 365 (1934).

22. K. Stewartson, *Proc. Camb. Phil. Soc.*, *49*, 333 (1953).

23. G. K. Batchelor, *Quart. J. Mech. Appl. Math.*, *4*, 29 (1951).

24. K. Liu and W. E. Stewart, *J. Heat Mass Transfer*, *15*, 187 (1972).

25. E. M. Sparrow and J. L. Gregg, *J. Heat Transfer, Trans. ASME*, *81*, 249 (1959).

26. M. H. Rogers and G. N. Lance, *J. Fluid Mech.*, *7*, 617 (1970).

27. P. Capper and D. Elwell, *J. Crystal Growth*, *30*, 352 (1975).

28. S. T. McComas and J. P. Hartnett, in *Heat Transfer 1970*, Vol. 3, (U. Griqull and E. Hahne, eds.), Elsevier, Amsterdam, 1970, paper no. FC7.7.

29. F. Kreith and H. Viviand, *J. Appl. Mech., Trans. ASME*, *89*, 541 (1967).

30. R. Hide and C. W. Titman, *J. Fluid Mech.*, *29*, 39 (1967).

31. C. E. Pearson, *J. Fluid Mech.*, *21*, 623 (1965).

32. E. M. Sparrow and J. L. Gregg, *J. Heat Transfer, Trans. ASME*, *82*, 294 (1960).

33. L. O. Wilson, *J. Crystal Growth*, *44*, 247 (1978).

34. L. O. Wilson, *J. Crystal Growth*, *44*, 371 (1978).

35. L. O. Wilson and N. L. Cahryer, *J. Fluid Mech.*, *85*, 479 (1978).

36. H. P. Greenspan, *The Theory of Rotating Fluids*, Cambridge University Press, Cambridge, 1969, chap. 2.

37. A. J. Goss and R. E. Adlington, *Marconi Rev.*, *22*, 18 (1959).

38. B. M. Turovskii and M. G. Mil'vidskii, *Sov. Phys.-Crystallog.*, *6*, 606 (1962).

39. D. S. Robertson, *Brit. J. Appl. Phys.*, *17*, 1047 (1966).

40. J. R. Carruthers, *J. Electrochem. Soc.*, *114*, 959 (1967).

41. J. R. Carruthers and K. Nassau, *J. Appl. Phys.*, *39*, 5205 (1968).

42. K. Shiroki, *J. Crystal Growth*, *40*, 129 (1977).

43. C. D. Brandle, *J. Crystal Growth*, *42*, 400 (1977).

44. N. Kobayashi and T. Arizumi, *J. Crystal Growth*, *49*, 419 (1980).

45. N. Kobayashi and T. Arizumi, *Jap. J. Appl. Phys.*, *9*, 361 (1970).

46. N. Kobayashi and T. Arizumi, *Jap. J. Appl. Phys.*, *9*, 1255 (1970).

47. N. Kobayashi and T. Arizumi, *J. Crystal Growth*, *30*, 177 (1975).

48. N. Kobayashi, *J. Crystal Growth*, *43*, 357 (1978).

49. N. Kobayashi, *Comp. Meths. Appl. Meth. Eng.*, *23*, 21 (1980).
     press.

50. W. E. Langlois, *Appl. Math. Modelling*, *1*, 196 (1977).

51. W. E. Langlois and C. C. Shir, *Comp. Meths. Appl. Mech. Eng.*, *12*,
     145 (1977).

52. W. E. Langlois, *J. Crystal Growth*, *42*, 386 (1977).

53. W. E. Langlois, *Proc. 6th Int. Congr. Num. Methods Fluid Dynamics*, Tbilisi, USSR, 1970.

54. W. E. Langlois, *J. Crystal Growth*, *46*, 743 (1979).

55. W. E. Langlois, *J. Crystal Growth*, *48*, 25 (1980).

56. J. R. Carruthers, *J. Crystal Growth*, *36*, 212 (1976).

57. J. R. Carruthers and A. F. Witt, in *Crystal Growth and Characterization* (R. Ueda and J. B. Mullin, eds.), North-Holland,
     Amsterdam, 1974, p. 107.

58. J. A. M. Dikhoff, *Solid-State Electronics*, *1*, 202 (1960).

59. F. D. Rosi, *RCA Rev.*, *19*, 349 (1959).

60. B. Cockayne, M. Chesswas, and D. B. Gasson, *J. Mater. Sci.*, *3*,
     224 (1968).

61. B. Cockayne, M. Chesswas, and D. B. Gasson, *J. Mater. Sci.*, *4*,
     450 (1969).

62. I. E. Drafall and K. E. Spear, *J. Crystal Growth*, *33*, 180 (1976).

63. S. Miyazawa and H. Iwasaki, *Jap. J. Appl. Phys.*, *9*, 441 (1970).

64. T. Arizumi and N. Kobayashi, *Jap. J. Appl. Phys.*, *8*, 1091 (1969).

65. T. Arizumi and N. Kobayashi, *J. Crystal Growth*, *13/14*, 615 (1972).

66. W. R. Wilcox, in *Fractional Solidification* (M. Zief and W. R.
     Wilcox, eds.), Marcel Dekker, New York, 1967, chap. 6.

67. T. Inoue, *Trans. JIM*, *17*, 227 (1976).

68. M. G. Mil'vidskii and B. I. Golovin, *Sov. Phys.-Solid State*, *3*,
     737 (1961).

69. J. C. Brice, *Acta Electronica*, *16*, 291 (1973).

70. L. R. Rothrock, *J. Crystal Growth*, *39*, 180 (1977).

71. Y. Furuhata, *J. Jap. Assoc. Crystal Growth* (in Japanese), *3*, 1
     (1976).

72. J. C. Brice and P. A. C. Whiffin, *Solid-State Electronics*, *7*,
     183 (1964).

73. C. H. Sworn and T. E. Brown, *J. Crystal Growth*, *15*, 195 (1972).

74. J. C. Brice, O. F. Hill, P. A. C. Whiffin, and J. A. Wilkinson,
     J. Crystal Growth, *10*, 133 (1971).

75. G. Zydzik, *Mater. Res. Bull.*, *10*, 701 (1975).

76. B. Cockayne, B. Lent, and J. M. Roslington, *J. Mater. Sci.*, *11*, 259 (1976).

77. K. Takagi, T. Fukazawa, and M. Ishii, *J. Crystal Growth*, *32*, 89 (1976).

78. D. C. Miller, A. J. Valentino, and L. K. Shick, *J. Crystal Growth*, *44*, 121 (1978).

79. Y. Miyazawa, Y. Mori, and S. Homma, *J. Crystal Growth*, *43*, 541 (1978).

80. Y. Miyazawa, Y. Mori, S. Homma, and K. Kitamura, *Mater. Res. Bull.*, *13*, 675 (1978).

81. J. C. Brice, T. M. Bruton, O. F. Hill, and P. A. C. Whiffin, *J. Crystal Growth*, *24/25*, 429 (1974).

82. J. C. Brice and P. A. C. Whiffin, *J. Crystal Growth*, *38*, 245 (1977).

83. P. A. C. Whiffin, T. M. Bruton, and J. C. Brice, *J. Crystal Growth*, *32*, 205 (1976).

84. H. P. Greenspan, *The Theory of Rotating Fluids*, Cambridge University Press, Cambridge, 1969, p. 293.

85. M. Kuriyama, J. C. Earley, and H. E. Burdette, AIAA Paper 74-204, 1 (1974).

86. D. S. Robertson and I. M. Young, *J. Phys.*, *D, Appl. Phys.*, *8*, 59 (1975).

87. B. Cockayne, *J. Crystal Growth*, *3/4*, 60 (1968).

88. B. Cockayne, J. G. Plant, and A. W. Vere, *J. Crystal Growth*, *15*, 167 (1972).

89. I. Shih and C. H. Champness, *J. Crystal Growth*, *44*, 492 (1978).

90. F. J. Bruni, in *Crystals for Magnetic Applications* (C. J. M. Rooijmans, ed.), Springer-Verlag, Berlin, 1978, p. 53.

91. J. R. Carruthers, in *Preparation and Properties of Solid State Materials*, Vol. 3 (R. A. Lefever and W. R. Wilcox, eds.), Marcel Dekker, New York, 1977, p. 1.

92. J. R. Carruthers, *J. Crystal Growth*, *32*, 13 (1976).

93. B. Cockayne, M. Chesswas, J. G. Plant, and A. W. Vere, *J. Mater. Sci.*, *4*, 565 (1969).

94. P. A. C. Whiffin and J. C. Brice, *J. Crystal Growth*, *10*, 91 (1971).

95. W. R. Wilcox and L. D. Fullmer, *J. Appl. Phys.*, *36*, 2201 (1965).

96. J. C. Brice, *J. Crystal Growth*, *42*, 427 (1977).

97. W. C. Dash, *J. Appl. Phys.*, *30*, 459 (1959).

98. S. V. Tsivinskii, *Inzh. Fiz. Zh.*, *5*, 59 (1962).

99.   T. Surek, *J. Appl. Phys.*, *47*, 4384 (1976).

100. E. R. G. Eckert and R. M. Drake, Jr., *Analysis of Heat and Mass Transfer*, McGraw-Hill Kogakusha, Tokyo, 1972, chap. 16.

101. R. Siegel and J. R. Howell, *Thermal Radiation Heat Transfer*, McGraw-Hill Kogakusha, Tokyo, 1972, chap. 19.

102. J. C. Brice and P. A. C. Whiffin, *Brit. J. Appl. Phys.*, *18*, 581 (1967).

103. J. C. Brice, *J. Crystal Growth*, *6*, 205 (1970).

104. T. Abe, *J. Crystal Growth*, *24/25*, 463 (1974).

105. K. A. Jackson, in *Liquid Metals and Solidification*, American Society for Metals, Cleveland, p. 174.

106. K. M. Kim and H. E. Temple, *J. Crystal Growth*, *34*, 177 (1976).

107. T. Fukuda and H. Hirano, *J. Crystal Growth*, *35*, 127 (1976).

108. J. B. Mullin, W. R. MacEvan, C. H. Holliday, and A. E. V. Webb, *J. Crystal Growth*, *13/14*, 629 (1972).

109. M. Chesswas, B. Cockayne, D. T. J. Hurle, E. Jakeman and J. B. Mullin, *J. Crystal Growth*, *11*, 225 (1971).

110. S. F. Nygen, *J. Crystal Growth*, *19*, 21 (1973).

111. E. M. Sparrow, G. B. Miller and V. K. Jonsson, *J. Aerospace Sci.*, *29*, 1291 (1962).

112. G. E. Forsythe and W. R. Wasaw, *Finite-Difference Methods for Partial Differential Equations*, Wiley, New York, (1960).

AUTHOR INDEX

Numbers in parentheses are reference numbers and indicate that an author's work is referred to although the name may not be cited in the text. Numbers in italics give the page on which the complete reference is cited.